the
# LIGHTDARK
## universe

**James M Overduin**
*Stanford University, USA*

**Paul S Wesson**
*University of Waterloo, Canada*

# the LIGHTDARK universe

## Light from Galaxies, Dark Matter and Dark Energy

 **World Scientific**

NEW JERSEY · LONDON · SINGAPORE · BEIJING · SHANGHAI · HONG KONG · TAIPEI · CHENNAI

*Published by*

World Scientific Publishing Co. Pte. Ltd.

5 Toh Tuck Link, Singapore 596224

*USA office:* 27 Warren Street, Suite 401-402, Hackensack, NJ 07601

*UK office:* 57 Shelton Street, Covent Garden, London WC2H 9HE

**British Library Cataloguing-in-Publication Data**
A catalogue record for this book is available from the British Library.

**THE LIGHT/DARK UNIVERSE**
**Galactic Light, Dark Matter and Dark Energy**

ISBN-13   978-981-283-441-6
ISBN-10   981-283-441-9
ISBN-13   978-981-283-589-5 (pbk)
ISBN-10   981-283-589-X

Printed in Singapore

# Preface

Why is the sky dark at night? If the Universe is infinite and uniformly populated with luminous galaxies which have existed forever, then the night sky should be ablaze with light. Obviously it is not—but why? Thinkers through the ages have come up with at least a dozen different answers to this question, which was dubbed "Olbers' paradox" in 1952. It was not really paradoxical then; nearly all of those who noticed the puzzle (including Olbers himself in 1823) were perfectly content with their answers, whether right or wrong. And it is neither paradoxical nor puzzling today: the reasons for the darkness of the night sky are now well understood. Yet the problem runs so deep, and touches upon so many fundamental aspects of cosmology and metaphysics, that it continues to hold a perennial fascination for astronomers and the public alike. Perhaps most profoundly of all, the darkness of the Universe at optical wavelengths is a clue to the *finiteness in time* of those sources of light that we call home: the stars and galaxies. They could not have existed forever, or the cosmos would have filled up with light. The fact that it has not tells us approximately how long they have been shining. In fact, by measuring the intensity of the night sky and applying some simple physics, we can estimate the elapsed time since the big bang with a fair degree of accuracy. Alternatively, we can calculate exactly *how* dark the sky should be at night, using what astronomers have learned about stars and galaxies together with the dynamics of the Universe according to Einstein's theory of general relativity. The results agree closely with what we see.

Such is the precision of modern observational cosmology, however, that we can go further than this and use the exact level of intensity of the extra-galactic background light at all wavelengths (not just the optical) to look for hints as to *what else* may or may not be shining in the Universe. Cosmologists are now convinced that the Universe is dynamically dominated by two mysterious and apparently independent substances, known as dark matter and dark energy, whose energy density dwarfs that of conventional matter and radiation, and whose properties are inconsistent with anything in the existing standard model of particle physics. Very little is known about these new forms of matter-energy. However, most of the candidates that have been proposed so far are not *completely* dark. Rather, they decay into or otherwise interact with photons in characteristic ways that can be accurately modelled and compared with observational data. Experimental limits on the intensity of cosmic background radiation in the microwave, infrared, optical, ultraviolet, x-ray and gamma-ray bands rule out certain kinds of decaying dark energy, as well as dark matter in the form of light axions, neutrinos, unstable weakly-interacting massive particles (WIMPs) and objects like black holes. Thus does Olbers' paradox gain new importance as a window on the Universe, both seen and unseen.

The topic of the dark night sky is one which we, as authors, have had the opportunity to study not only as a pastime but also as a profession. We are grateful for the input of numerous researchers, and for the hospitality of several universities, notably Berkeley and Stanford. However, as we emphasize in the following (mainly technical) account, *anybody* with a clear mind can make the connection between the fact that the night sky is still dark and the fact that the Universe is young. The naked eye confirms the birth of the Universe in something like a big bang. The poet and writer Edgar Allan Poe guessed as much when he made the connection between darkness and age long before there were astrophysicists as such. We therefore dedicate this book to the thinking reader, who has looked at the glory of the night sky and wondered *what does it mean?*

*J.M. Overduin and P.S. Wesson*

# Contents

Chapter 1

# The Enigma of the Dark Night Sky

## 1.1 Why is the sky dark at night?

Most of us find nothing remarkable in the fact that the sky is dark at night. Yet remarkable it certainly is. The Universe is filled with galaxies like our own Milky Way. They are far away, to be sure, and their apparent luminosity falls off as distance squared. Thus galaxies located inside an imaginary spherical shell ten times farther away than our nearest large neighbor, Andromeda, appear a hundred times fainter, and galaxies beyond that are dimmer still. However, this decrease in brightness is made up by an increase in the *number* of galaxies we can see at the same distance, because the surface area of the spherical shell that contains them *goes up* as distance squared. Every such shell, in fact, adds the same amount of light to the brightness of the night sky. And as we look farther and farther out, we take in more and more of these shells, until at last every line of sight must end on a remote galaxy, much as every line of sight in a large forest ends up on a tree (Fig. 1.1).

If we have described the situation correctly, then the view in all directions must ultimately be blocked by overlapping galaxies, and the light from these, propagating towards us from the distant reaches of the Universe, should make the night sky look as bright as the surface of a star—brighter even than the starry firmament over Arles as painted by Vincent Van Gogh in 1889 (Fig. 1.2).

Clearly this does not happen in reality. The largest modern telescopes, trained on the same region of sky for hours, do find numerous galaxies (Fig. 1.3), but nothing like a continuous sea of light. Why not? Since H. Bondi's book *Cosmology* in 1952 [1] this question has become inextricably linked with the name of H.W.M. Olbers, a German physician and astronomer who published an account of the problem in 1823 [2].

1

Fig. 1.1   In a large forest of evenly spaced trees, every line of sight must eventually end up at a tree.

Olbers was not the first to mention it; historians and scientists have identified several who preceded him, and many others who came tantalizingly close in the two centuries before. In fact it is interesting that the puzzle did not come to light two *millenia* earlier than it did. The essential facts were already in hand in Greek antiquity. (The above argument is unchanged if we substitute "stars" for "galaxies.") It nearly came within the grasp of the presocratic philosopher Democritus (c. 460-370 BCE), who recognized that the fuzzy band of light we know as the Milky Way is in reality "composed of very small, tightly packed stars which seem to us joined together because of the distance of the heavens from the Earth, just as if many fine grains of salt had been poured out in one place" [4]. But despite the fact that Democritus (along with the other Greek atomists) believed in an infinite Universe, and recognized the existence of other stars besides the Sun, he did not carry the argument farther and draw the logical conclusion that the sky in such a Universe should resemble the Milky Way *in every direction*. Nor is there any record that the atomists' philosophical rivals, the Stoics, seized upon the darkness of the night sky as evidence for their own alternative theory of a finite cosmos.

Fig. 1.2   Vincent Van Gogh's "Starry Night" (1889).

The history of what E.R. Harrison has called the "riddle of cosmic darkness" [5] is a tangled and fascinating one that has been studied by many people [3–18][1] In the remainder of this chapter, we will identify at least a dozen different ways in which various thinkers have proposed to answer the question "Why is the sky dark at night?" Some of their answers strike the modern mind as muddled, mistaken, or both; but others have proved prescient, and some have persisted in cosmology to this day, albeit not always in their original form. Such is the fundamental nature of the problem, which D.W. Sciama hailed as "the first discovery of a link connecting us to the distant regions of the Universe" [19], that these ideas still retain remarkable suggestive power today.

---

[1] We are particularly indebted to Refs. [4] and [5] in what follows; quotations are drawn from these two invaluable sources except where otherwise noted.

Fig. 1.3  View of the distant Universe as seen by one of the most powerful eyes on Earth: the giant MegaCam camera mounted at the prime focus of the 3.6 meter Canada-France-Hawaii Telescope. This closeup shows less than 1% of a single 1 degree-squared MegaCam field in one of the darkest regions of the sky (in the constellation Cetus)—about the same area that you might see while peering through a long straw. Apart from a handful of foreground stars, *every one of the thousands of objects seen here is a galaxy.* (Image courtesy of Henry J. McCracken, CFHT/CNRS/IAP/Terapix.)

## 1.2   "By reason of distance"

First to allude to the problem, albeit indirectly, was the English astronomer Thomas Digges. In 1576, inspired with zeal for the Copernicus' new Earth-centered model, he was moved to add to a book written by his father (also an astronomer) an appendix titled "Perfit Description of the Caelestiall Orbes according to the most aunciente doctrine of the Pythagoreans latelye reuiued by Copernicus and by Geometricall Demonstrations approued." In it he described for the first time an unbounded Universe whose stars were no longer fastened to a celestial sphere (as imagined by Aristotle and Ptolemy), but rather "lights innumerable and reaching vp in Sphaericall altitude without ende ... and as they are hygher, so seeme they of lesse and lesser quantitye, euen tyll our sighte beinge not able farder to reache

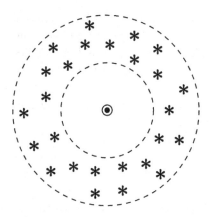

Fig. 1.4 Kepler's island Universe (1610). The sky looks dark at night because the stars extend only out to a finite distance, the "edge of the Universe."

or conceyue, the greatest part rest by reason of their wonderfull distance inuisible vnto vs."

Here was not only an infinity of stars, but also the first dim recognition of the need to explain why it did not *look* infinite. Digges' vague reference to "distance" was not fully thought out. However, explanations in a similar vein continued to be offered for many years by eminent scientists including William Gilbert (1600) and Galileo Galilei (1624). In modern terms, Digges' phrase "by reason of distance" might be interpreted in either of two ways: as a suggestion that the intensity of light drops off more rapidly than the inverse square of the distance over cosmological distances ("tired light"), or that distant sources themselves are *intrinsically* less bright than nearby ones ("non-uniform sources"). Both these ideas would later appear in more specific and interesting forms.

## 1.3 Island Universe

Johannes Kepler was the first to comment on the problem directly, writing in 1610 that "in an infinite Universe the stars would fill the heavens as they are seen by us" [18]. If he did not view this possibility with more urgency than Digges, it was because Kepler was perfectly content with a bounded Universe containing a *finite* number of stars (Fig. 1.4). When he received a copy of Galileo's *Starry Message* in 1610, with its report of many new stars visible through the telescope, he replied: "You do not hesitate to declare that there are visible over 10,000 stars ... If they are suns having the same nature as our Sun, why do not these suns collectively outdistance our Sun in brilliance? Why do they all together transmit so dim a light to the most accessible places?" So firm was Kepler's belief in a finite cosmos that such observations only suggested to him that the new stars must be intrinsically fainter than the Sun. Had he been alarmed enough to take his own argument more seriously, we would probably speak today of "Kepler's paradox" rather than Olbers'.

Kepler's island Universe, "enclosed and circumscribed as by a wall or a vault," may seem quaintly old-fashioned. Remove the wall, however, and it becomes the prototype for many a finite cosmology to come. Kepler's countryman Otto von Guericke, who was the first to compare the sky of stars to a forest of trees, advocated an island Universe floating alone in an infinite void resembling that of the Stoic philosophers. He guessed in 1672 that "probably no more stars will be seen shining after passing through the whole stellar realm made visible with the telescope." The Dutch scientist Christiaan Huygens found a moral in the finitude of the island Universe (1698): "What if beyond such a determinate space [God] has left an infinite Vacuum to show how inconsiderable is all that he has made, to what his Power could, had he so pleas'd, have produc'd?"

Finite cosmological models were not completely discounted as possible explanations for the dark night sky until the 1920s and 1930s, when the existence and spatial extent of the extragalactic Universe were definitively established by observational astronomy, and the notion of large-scale homogeneity achieved scientific legitimacy in the context of cosmological models based on Einstein's theory of general relativity. Thus, as noted by H. Kragh [18], it was still possible as late as 1917 to find an astronomer of the stature of H. Shapley appealing to Kepler's resolution of Olbers' paradox in support of the view that there were no stars outside the Milky Way: "Either the extent of the star-populated space is finite or 'the heavens would be a blazing glory of light' ... Since the heavens are not a blazing glory, and since space absorption is of little moment ... it follows that the defined stellar system is finite."

## 1.4  Non-uniform sources

The person who is widely credited with introducing the first clear formulation of the paradox is Edmond Halley, who wrote in 1721: "I have heard urged, that *if the number of Fixt Stars were more than finite, [then] the whole superficies of their apparent Sphere would be luminous"* ([20], emphasis ours). This is explicit enough to show that Halley fully grasped the mystery of the dark night sky. However historians have long wondered exactly what he meant by "luminous," and who did the urging. A nice piece of detective work by M. Hoskin [12] has cleared up both mysteries.

The story goes back to Isaac Newton, whose new theory of gravity (1687) seemed to require an infinite number of stars, spaced evenly apart in an infinite Universe. If the spacing were not perfectly even, and did not extend to infinity, then each star would feel slightly more gravitational force in one direction than another, and the entire assembly would collapse in upon itself like a house of cards. Newton himself recognized that such an arrangement was inherently unstable, and confided to a student that a "continual miracle" was needed to prop it up. However it was the optical consequences of the model that bothered a young physician named William Stukeley, who raised the following objection with Newton around 1720: "What w[oul]d have been the consequence had infinite space ... been disseminated with worlds? We see every night, the inconvenience of it. *The whole hemisphere w[oul]d have had the appearance of that luminous gloom of the Milky Way"* (emphasis ours). Stukeley's memoirs report that he discussed astronomical matters with Newton and Halley over breakfast just two weeks before Halley read his paper before the Royal Society in 1721 [16]. The identity of the anonymous "urger" and probable originator of what we know as Olbers' paradox is thus revealed.

There is no record of Newton's response. However, in 1692 he had faced a similar problem while working on an unpublished revision of his *Principia.* The handwritten manuscript shows that Newton had decided to test his theory of evenly spaced stars in an infinite Universe by comparing stellar brightnesses in such a model against observations in an actual star catalog. He arranged the stars in concentric spherical shells around the Sun in such a way that every star was equidistant from all its neighbors; this led to the prediction that there should be 12 or 13 stars in the first shell, $2^2 = 4$ times as many in the second, $3^2 = 9$ times as many in the third, and so on. Assuming at first that stellar magnitude was proportional to distance, he found to his satisfaction that these numbers agreed roughly with the number of first-, second- and third-magnitude stars in the catalog. Farther out, however, his model began to diverge badly from reality. Newton's response was to *adjust the relationship between brightness and distance,* "stretching" the distance to a star of given magnitude and thus obtaining better agreement with the observed numbers of fainter stars [17]. In the

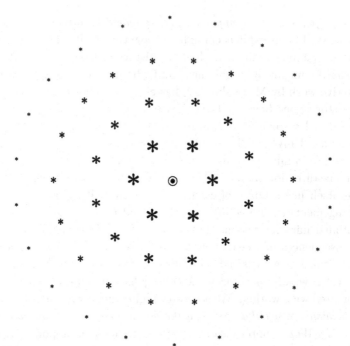

Fig. 1.5  Non-uniform sources. The sky might look dark at night if the intrinsic luminosity of the stars decreased with distance from the Sun, as considered by Kepler (1610) and possibly also by Newton (1692).

process he explicitly noted that the stars at two units of distance were four times as numerous—and four times fainter—than those at one [8]. But in one of the great near misses of science, Newton failed to notice that each of his shells would then be equally bright, and hence that all the shells together would fill the sky with light.

Newton was concerned with matching up the numbers of stars in his theory to those in the star catalog, not with explaining why the sky is dark at night (a puzzle he did not notice). Unlike Kepler, he probably intended to alter only the way in which the *apparent* brightnesses of the stars depended upon their distance from the Sun (not their absolute, or intrinsic luminosities, which he presumably took to be as uniform as their masses). Nevertheless this episode raises the fascinating possibility of a different resolution to Olbers' paradox, namely to suppose that the sources of light in the Universe are not uniformly luminous but rather *fade intrinsically with distance* (Fig. 1.5). So profoundly anti-Copernican a proposal would be

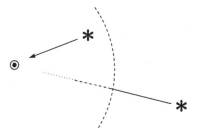

Fig. 1.6 Tired light. N. Hartsoeker (1707) suggested that the sky is dark at night because the intensity of light drops off more quickly than expected over cosmological distances. A similar explanation was offered by E. Halley in 1721.

difficult for us to accept today. However, we will find in Sec. 4.6 that similar ideas are being raised in the context of modern cosmology as a way of explaining the apparent dimming of distant supernovae.

## 1.5 Tired light

If Kepler and Newton provide us with examples of ways in which one might attempt to account for a dark night sky by altering the properties of sources, then the alternative—altering the properties of *light*—might be traced back to Newton's rival Réné Descartes. His "vortex theory" of mechanics (as described in the *Principles of Philosophy*, 1644) included a provision for the weakening of light as it passed from the vortex surrounding one star to another. But the precise way in which this happened was not well explained, and Descartes did not explicitly connect it with the darkness of the night sky, even though he postulated an infinite Universe populated by innumerable stars like the Sun.

Such a connection was however made explicitly by the Dutch scientist Nicolaas Hartsoeker, who wrote about the stars in a set of lectures titled *Physical Conjectures* (1707): "Their number is infinite. From this it follows, my Lord, that *the rays of light must diminish and disappear on their route to us, or else the whole sky would be as luminous as the Sun*" (Fig. 1.6, emphasis ours). As S.L. Jaki [4] has written, "The optical paradox ... was here stated in perfect clarity and conciseness, together with a solution to it, namely the diminution of starlight through cosmic spaces. Only the cause of that diminution failed to be given." Similar comments apply to Halley, who attempted to resolve the paradox in 1721 by arguing that "when the stars are at very remote distances their light diminishes in a greater proportion than according to the common rule and at last becomes entirely insensible even to the largest telescopes" [8].

There have been attempts to put such a diminution on a physical footing in modern times, most notably in the form of the "tired light" hypothesis of F. Zwicky and others in the late 1920s and 1930s. These efforts, however, have typically been motivated, not by the darkness of the night sky, but rather by a desire to account for cosmological redshift without expansion. (In the former context, they face the same challenge as Olbers' own solution, to be discussed in the next section.) While they have been resurrected from time to time in more sophisticated forms (e.g., by E. Finlay-Freundlich and M. Born in 1953), tired-light theories have not survived observational test.

## 1.6   Absorption

The best-known (incorrect) explanation for the darkness of the night sky is undoubtedly the one proposed by Olbers himself; namely, that light rays from distant stars are absorbed en route to us by an interstellar medium. Prescientific roots for such an idea can perhaps be discerned in the speculations of Thomas Campanella, who suggested in 1620 that most of the infinitely many stars are invisible because their light is blocked by a thick layer of vapor created by the "continual clash between the holiness of the heavens and the depravity of the Earth" [4].

The first person to couch this solution in scientific terms was not Olbers, but the Swiss astronomer Jean-Philippe Loys de Chéseaux, who wrote in 1744 [21]: "Either ... the sphere of fixed stars is not infinite but actually incomparably smaller than the finite extension I have supposed for it, or ... the force of light decreases faster than the inverse square of distance. This latter supposition is quite plausible; *it requires only that starry space is filled with a fluid capable of intercepting light very slightly*" (Fig. 1.7, emphasis ours). Olbers largely echoed that reasoning eighty years later in 1823 [2], adding that the assumption that interstellar space is perfectly transparent to light is "not only undemonstrated but also highly improbable." For him absorption was evidence of the benevolent wisdom of God, who "created a Universe of great yet not quite perfect transparency, and has thereby restricted the range of vision to a limited part of infinite space. Thus we are permitted to discover some of the design and construction of the Universe, of which we could know almost nothing if the remotest suns were allowed to blaze with undiminished light."

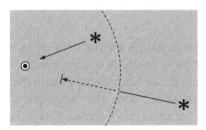

Fig. 1.7 Absorption. J.-P.L. De Chéseaux (1744) and H.W.M. Olbers (1823) were the first to propose that the sky is dark at night because starlight is absorbed in an interstellar medium.

The fact that what we know as a "paradox" has become so closely associated with Olbers' name is one of the story's most paradoxical aspects, since he explicitly quoted Halley's account of the problem from 1721, and less explicitly borrowed much of his resolution from de Chéseaux. Olbers owned a copy of de Chéseaux's book, and since it dealt with comets, the field in which both men did most of their work, he had almost certainly read it. Many years had passed, however, and since Olbers was known as a man of scrupulous integrity the historical consensus [3, 7, 15] is that he simply forgot about his predecessor.

The question is of largely academic interest in any case, since absorption fails to solve the paradox on quite general thermodynamic grounds. This point was first made in 1848 by John Herschel, who noted that light, "though absorbed, remains still effective in heating the absorbing medium, which must either increase in temperature ... or, in its turn becoming radiant, give out from every point at every instant as much heat as it receives." In other words, conservation of energy requires that any such medium must eventually come into thermodynamic equilibrium with the starlight, at which point it will be glowing as brightly as the stars themselves. Thus the paradox is not averted but merely *delayed*, which in an infinitely old Universe is no solution at all.

If the stars and galaxies are *not* infinitely old, and in particular if they are younger than the timescale required to reach thermodynamic equilibrium with the intergalactic medium, then absorption can still play an important—though not dominant—role in darkening the night sky. The energy given off by the stars cannot vanish, but can be "reprocessed" by such a medium. We will find in Chap. 3 that intergalactic gas and dust are able to shift a significant fraction of visible starlight into the infrared part of the electromagnetic spectrum.

## 1.7   Fractal Universe

An entirely different explanation for the dark night sky was hinted at by
Herschel in the same 1848 publication mentioned above: namely, to dis-
tribute the light sources in the Universe in a hierarchical or *fractal* manner,
or what he referred to as following the "principle of subordinate group-
ing," so that stars are grouped into galaxies, galaxies into clusters, clusters
into super-clusters, and so on, literally without end. The idea for such a
"scale-invariant" cosmos goes back at least to the German mathematician
Johann Heinrich Lambert and the philosopher Immanuel Kant. The latter
wrote as follows in his *Universal Natural History and Theory of the Heavens*
(1755): "... all these immense orders of star-worlds again form but one of
a number whose termination we do not know, and which perhaps, like the
former, is a system inconceivably vast—and yet again but one member in
a new combination of numbers! ... There is here no end but an abyss of
real immensity, in presence of which all the capability of human conception
sinks exhausted."

In such a Universe, the mean density of sources decreases as it is aver-
aged over progressively larger volumes, and tends toward zero in the infinite limit. The result, as Herschel stressed in an 1869 letter to the English
astronomer Richard Proctor, is "a Universe literally infinite which would
allow of any amount of such directions of penetration as not to encounter
a star" (Fig. 1.8). Proctor had come up with the same idea independently
and described it in his semi-popular book *Other Worlds Than Ours* (1871).
It was also colorfully described in *Two New Worlds* by the British scientist
and engineer E.E. Fournier d'Albe (1907). The latter account greatly im-
pressed Swedish astronomer C.V.L. Charlier, who worked out the general
conditions for transparency in a fractal Universe together with his German
counterpart H. von Seeliger between 1908 and 1922.

Hierarchical models have continued to attract attention down to the
present day from notable thinkers such as B. Mandelbrot, but have never
gained a strong foothold in cosmology. They contain no natural place for
phenomena such as Hubble's law, microwave background radiation or primordial nucleosynthesis, all of which fit together neatly in the context of
relativistic big-bang cosmology. Their most immediate drawback, however,
is conflict with observations like those in Fig. 1.3, which reveal that the
Universe on scales larger than galaxy clusters is not hierarchical but homogeneous. In order for the fractal Universe to work as an explanation for
the dark night sky, the hierarchy of clustering scales must *literally* extend

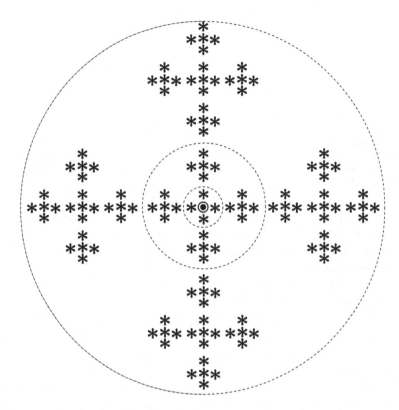

Fig. 1.8   Fractal Universe. J. Herschel (1848), R. Proctor (1871), E.E. Fournier-d'Albe (1907) and C.V.L. Charlier (1908) argued that the night sky would be dark if the stars were distributed in a hierarchical manner. At each size scale in the hierarchy (dashed circles) the distribution of stars looks the same.

to infinity. If there is an upper limit to the hierarchy at (say) the galaxy-cluster scale, then "clusters" merely take the place of "stars" and Olbers' paradox reappears in full force.

## 1.8   Finite age

Two pieces of information had to come together before it was possible to understand why the sky is really dark at night. The first of these pieces is the finite speed of light, whose existence was suspected by the Italian astronomer Gian Cassini in 1669 and demonstrated by his Danish counterpart Ole Roemer in 1676 using observations of eclipses of the moons of

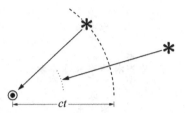

Fig. 1.9   Light horizon. If stars came into existence at a time $t$ in the past, then we cannot see them at distances greater than $ct$ where $c$ is the finite speed of light (dashed line). This fact seems to have been first glimpsed by E.A. Poe (1848) and then stated clearly by J.H. Mädler (1858).

Jupiter. If light travels at a finite speed, then it follows immediately that the light we receive from distant stars and galaxies *was emitted in the distant past*. Astoundingly, the first person to connect this fact to the light of the night sky was not a scientist but the American poet Edgar Allan Poe, who argued as follows in *Eureka: A Prose Poem* (1848): "Were the succession of stars endless, then the background of the sky would present us an uniform luminosity, like that displayed by the Galaxy—since there could be absolutely no point, in all that background, at which would not exist a star. The only mode, therefore, in which ... we could comprehend the voids which our telescopes find in innumerable directions, would be *by supposing the distance of the invisible background so immense that no ray from it has yet been able to reach us at all*" (Fig. 1.9, emphasis ours). The physicist E.R. Harrison has cited this as the "first clear and correct solution to the riddle of darkness, though only qualitatively expressed."

However, light-travel time is only half the story. If the stars have been shining forever, then it hardly matters how far away they are—their light has had an infinitely long time to reach us, and the riddle of darkness remains. Thus Poe, who advocated an infinitely old Universe (albeit one undergoing periodic conflagrations), did not really resolve the problem, as emphasized by F.J. Tipler [14]. One more thing is needed besides the finite speed of light: the *finite age of the stars*. If the stars came into existence at a time $t$ in the past, and their light travels with speed $c$, then it follows that they cannot be seen beyond a distance $d = ct$, because they *had not yet come into being at that point*. There is, in other words, an imaginary spherical surface around us—a light horizon—which defines the edge of the observable Universe. The light of our night sky comes only from the limited number of galaxies within this horizon, and even if we could peer further out we would only see the presumably non-luminous material from which they formed. This is the main reason why the sky is dark at night, as was first clearly understood by the German astronomer Johann H. Mädler in 1858 [22]: "The world is created, and hence is not eternal ... In the finite amount of time it could travel before it reached our eye, a light beam could pass through only a finite space, no matter how large the speed of light. *If we knew the moment of creation, we would be able to establish its boundary*" (emphasis ours). Today, we apply Mädler's basic suggestion in reverse: we use observations of the "boundary" where primordial radiation was generated to calculate the age of the Universe.

Why did it take so long to put finite lightspeed and finite lifetime together and arrive at this simple explanation of the dark night sky? Historians and scientists have long debated this, the "only remaining puzzling aspect of the dark night-sky riddle" [11]. Recall that the speed of light was measured in 1676, and that belief in a young Universe was widespread long before that (as witnessed by most religious creation stories). It is therefore truly remarkable that this puzzle provoked mostly silence and false leads for nearly two centuries, while giants like Hooke, Newton, Euler, Lagrange, Herschel, Laplace, Gauss, Humboldt and others strode the stage of science. Jaki [4] and Hoskin [16] have noted that many of these thinkers were more occupied with the mechanical movements of the cosmic clockwork than its optical consequences, but do not speculate on why that should have been the case. Harrison [5] has suggested that the traditional geometrical argument introduced by Halley, whereby the Universe is divided into spherical shells centered on the Earth, may have encouraged an overly static way of thinking about the problem that ignored the way light actually travels. He has also raised the intriguing possibility that the notion of light-travel time may have been avoided, even by leading scientists, because it implied a Universe that was unacceptably old by theological standards. Ironically, religious leaders of the time could have nudged astronomers toward the correct resolution of Olbers' paradox. Indeed, the obvious solution was the biblical one, that the Universe was very young!

Alternatively, as conjectured by Tipler [14], it may simply be that physics was insufficiently developed before the end of the 1850s for the *necessity* of the finite-age resolution to be appreciated. Hoskin [8] and Kragh [18] have emphasized the fact that what we call Olbers' "paradox" was never actually considered paradoxical (nor even particularly interesting) until about 1950. Those who did connect the appearance of the sky "down here" with abstract speculations about the Universe "out there" were largely satisfied with the solutions available to them, particularly Ke-

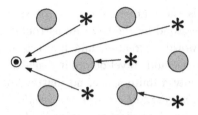

Fig. 1.10   Dark stars. F. Arago (1857) and E.E. Fournier d'Albe (1907) suggested that the sky may be dark at night because the light of distant stars is blocked by solid bodies.

pler's island Universe and de Chéseaux's absorbing medium. Elegant as it was, the light-horizon concept still required *two* ideas to come together at once. Perhaps it had to wait in the wings until simpler explanations based on a single idea proved demonstrably false.

## 1.9   Dark stars

Correct as it was, the finite-age resolution of Poe and Mädler languished in obscurity for over a century. (Even Mädler himself lost his enthusiasm for the idea, referring to it only as "another possibility" in his history of astronomy published 15 years later [14].) Meanwhile there was no shortage of further explanations for the darkness of the night sky. One such proposal, closely related to that of Olbers and de Chéseaux, was that starlight is blocked by large numbers of "dark stars," solid but nonluminous objects that might evade the thermodynamical objections to a diffuse medium, since their "equilibrium timescale" in the sense discussed above would far exceed any conceivable stellar lifetime (Fig. 1.10). Unseen heavenly bodies have long been staples of the cosmological imagination, figuring in the writings of Giordano Bruno (1584), Thomas Wright (1750) and the already-mentioned Johann Lambert. The last asserted in his *Cosmological Letters on the Arrangement of the World-Edifice* (1761) that luminous matter on each rung in the ever-ascending hierarchy of clustering scales was held in place by the gravitational attraction of massive "dark rulers." The possibility that such "completely obscure and opaque" bodies might play a role in resolving Olbers' paradox was first emphasized by the director of the Paris Observatory, F. Arago, in his posthumous *Popular Astronomy* (1857). The already-mentioned Fournier d'Albe took it seriously enough in *Two New Worlds* (1907) that he worked out the numerical ratio of required "black stars" relative to bright ones: 1000 to one [6].

Fanciful as these suggestions may sound, they are not too far removed from some of the exotic denizens of modern cosmology, like brown and red dwarf stars, black holes and dark matter. In fact, Fournier d'Albe's ratio is similar to current estimates for the ratio of densities of dark matter plus dark energy to luminous matter: approximately 250 to one (see Chap. 4). However it is extremely unlikely that dark stars could play a significant role in resolving Olbers' paradox. They could not consist of ordinary (baryonic) matter in any form, including ice, rocks, planets, stars or collapsed objects like black holes, since nucleosynthesis arguments constrain the *total* cosmic density of baryonic matter to only $8 \pm 6$ times that of the stars themselves (Sec. 4.2). Large objects of *any* composition are further constrained by general arguments based on phenomena like gravitational lensing, tidal distortion, clustering and structure formation. A smooth "fluid" composed of dark matter or dark energy in one of the forms discussed in Chaps. 5 to 9 might be considered. But such a medium would still need to conserve energy, and would need to couple to electromagnetic radiation in an artificial way in order to absorb starlight efficiently without causing conflict with observations in other wavebands.

## 1.10  Curvature

An ingenious and entirely different explanation for the dark night sky was proposed in 1872 by the German astrophysicist J.K.F. Zöllner. The essential idea was that the three dimensions of space are "positively curved" on large scales, so that parallel lines eventually meet rather than going on forever as in Euclidean geometry. A Universe of this type is unbounded but still finite in extent, thus limiting the number of light sources it can contain. In principle, one can look all the way around it and see the same star on opposite sides of the sky (Fig. 1.11). In practice this idea does not work, as discussed by Harrison [5, 24]. Light rays traveling "around the Universe" do not return exactly to their sources but are rather gravitationally deflected as they pass near massive objects. This process effectively "defocuses" the background light so that one *does* see starlight in every direction, no matter how curled up the Universe may be—unless the stars *have not existed forever*. In other words, the night sky in such a Universe is only dark if the galaxies are young, just as in one that is spatially flat.

The idea that space might be curved was not unique to Zöllner; it was studied at about the same time by the Canadian astronomer S. Newcomb

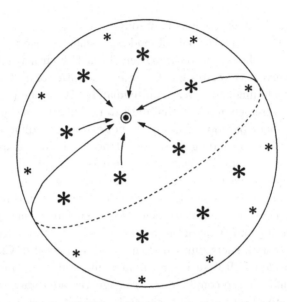

Fig. 1.11   Curvature.   J.K.F. Zöllner (1872) suggested that the sky is dark at night because space is positively curved and contains only a finite number of stars. In such a Universe it might be possible to see the same star on opposite sides of the sky (as indicated for one star).

(1877) and goes back to the great German mathematician C.F. Gauss as early as 1817. However, Zöllner was the first to apply non-Euclidean geometry to *cosmology*, as noted by Kragh [18]. He did so more than 40 years before general relativity—a prime example of the suggestive power of the dark-sky riddle.

Similar models also exist in Einstein's theory, and were discussed by the Dutch astronomer W. de Sitter in 1916 and 1917. He came within a hair's breadth of rediscovering both Olbers' paradox and Mädler's resolution, noting that in such a Universe we would "see an image of the back side of the Sun" in the antisolar direction unless light was absorbed in interstellar space, or unless the light-travel time around the Universe exceeded the age of the Sun. Einstein's static universe, one of two dominant cosmological models between 1917 and about 1930, was also of this type. There is no record that Einstein was motivated by the darkness of the night sky. However, his friend and colleague M. Born did explicitly draw that connection in the first edition of his pioneering relativity textbook of 1920 [25], suggesting (incorrectly) that the finiteness of Einstein's static model could explain why we do not see "the heavens ... shine with a bright light."

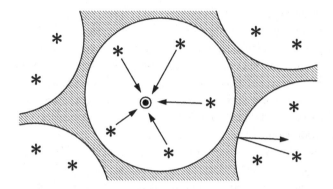

Fig. 1.12   Ether voids. S. Newcomb (1878) and J.E. Gore (1888) suggested that the sky is dark at night because our region of the Universe is surrounded by a hole in the ether (shaded area) through which light waves cannot pass. Light rays striking the edge of the hole may be reflected back (as shown at lower right).

More recently, the idea of curved spaces has re-entered cosmology in the form of objects known as strings, branes and solitons in higher dimensions. Thus do old attempts to come to grips with Olbers' paradox find new life in modern physics, even when wrong in their original contexts. We will discuss the possible consequences of some of these objects for the brightness of the night sky in Chap. 9.

## 1.11   Ether voids

In 1878, the above-mentioned astronomer S. Newcomb put forward another resolution to Olbers' paradox that depended on the now-discredited idea of a "luminiferous ether," a medium that was once thought necessary for the propagation of light. If this medium were not smoothly distributed, but rather punctuated by "ether voids," then light rays would effectively be blocked at their boundaries and one might in principle have another explanation for the dark night sky. This argument was elaborated by the Irish engineer J.E. Gore, who speculated that the region of the Universe containing the Milky Way might be *entirely surrounded* by such an ether void. In that case, he wrote in 1888, we would receive no light at all from external galaxies; only rays originating inside our own Milky Way would reach our eyes, either directly or indirectly after reflection from the "verge of the vacuum" (Fig. 1.12).

As noted by Harrison [5], such an idea fails on elementary grounds of energy conservation even if there *were* an ether. All that light bouncing back and forth between what are effectively reflecting walls around the Milky Way would eventually contribute as much to the brightness of the night sky as the light that would otherwise reach us from external galaxies. The only way to avoid this conclusion is to suppose that the stars have only been shining for a short time relative to the light-travel time between the edges of the void. Thus one is again driven back to Poe and Mädler's finite-age resolution.

While the idea of an ether void around the Milky Way would seem ludicrous today, it attracted interest from no less than W. Thomson (then Lord Kelvin) as late as 1902—though ultimately he found it improbable [5]. Modern cosmologists are not above raising ideas that are similar in certain respects, as evidenced by recent proposals to explain the unexpected behavior of the cosmological expansion rate by supposing that we are located near the center of a "local Hubble void" (see Chap. 4).

## 1.12   Insufficient energy

Kelvin's major contribution to the story of the dark night sky was to recast the problem and its correct solution in a new and clearer light, during a famous series of lectures in Baltimore in 1884. Unfortunately the article that resulted, first published in 1901 [23], was omitted in his official bibliographies, and fell into the same obscurity as the contributions of Poe and Mädler. Its importance has only recently been recognized [5]. What Kelvin did was simply to note that the time taken for light to reach us from the farthest visible stars is at least a million times longer than the lifetime of the Sun. From this it followed immediately that "if all the stars ... commenced shining at the same time, ... [then] at no instant would light be reaching the Earth from more than an excessively small proportion of all the stars." Here, in a nutshell, is the reason the sky is dark at night: *there is not enough energy in starlight, integrated over the history of the Universe, to make it bright.*

This is essentially the resolution glimpsed by Poe and, more explicitly, by Mädler. But it is clearer in its emphasis on the finite lifetime of the stars, which Kelvin was perhaps the first to appreciate in a thermodynamically rigorous way. In fact, he pushed his argument a step further, noting that one could "make the whole sky aglow with the light of all the stars at the same time"—but only by arranging for concentric spheres of stars to light up one after another like bulbs in a flashing neon sign, according to their nearness to the Earth! Such a cosmic conspiracy is obviously vanishingly improbable. Kelvin famously declared that "paradoxes have no place in science." As noted by Harrison, "It seems historically ironic that he was the first to solve with rigor and utmost lucidity a riddle that later, when his work lay forgotten, became known as Olbers' paradox."

By combining Kelvin's arguments with a little twentieth-century physics, Harrison has expressed the insufficient-energy resolution in a particularly compact and persuasive form [26–28]. Imagine that the entire rest mass of luminous matter in the Universe were converted directly into light. Recalling that visible matter in the Universe has an average density of about one proton per cubic meter, and applying Einstein's famous equation $E = mc^2$, one finds that the energy density of the resulting light is only equivalent to the energy density of *moonlight as received on Earth*. Moreover, stars do not convert their entire rest mass to light, but only about 0.1% of it. Thus, even in principle, the stars would not be capable of producing a night sky brighter than 1/1000 of that due to the Moon. Implicit in the traditional line-of-sight arguments of Halley, de Chéseaux and Olbers was an assumption that stars could shine forever, effectively converting their entire rest masses into light infinitely many times over. That, of course, is incompatible with energy conservation. Seen in this form, it seems inescapable that, whatever other factors may be at play, a major part of the the resolution of Olbers' paradox must lie in the simple fact that the stars have not shone long enough or brightly enough to fill the sky with light.

## 1.13 Light-matter interconversion

Two more false trails remained to be followed before Kelvin's thermodynamic insight re-emerged from obscurity. Both these trails were attempts to describe the Universe as a "steady state"—one that looks the same at all times as well as places—and both followed the Einsteinian revolution of 1916. However, one looked back to pre-relativistic cosmology, while the other was perhaps a little *too* relativistic, as we will see. The pre-relativistic steady-state theory came from the American astronomer W. MacMillan, who advocated a return to an infinite Euclidean Universe. To account for the darkness of the night sky in such a model, MacMillan postulated in 1918 that light rays from distant stars are *converted into atoms of mat-*

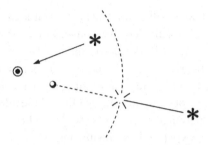

Fig. 1.13   Light-matter conversion. Macmillan (1918) suggested that the sky is dark at night because light from distant stars is converted into atoms of matter en route to us.

*ter* in such a way as to conserve energy (Fig. 1.13). The reverse reaction also occurs, he said, fulfilling the requirement that the appearance of the Universe as a whole does not change with time.

This proposal was met with skepticism, for many reasons. Perhaps most fundamentally, as emphasized by Harrison, thermodynamics does not consist only in the conservation of energy, but also in the inexorable increase of entropy. Energy cannot be perpetually "recycled" from one concentrated form into another and back; it inevitably disperses into forms where less work can be extracted from it.

Nevertheless light and matter *do* convert into each other in modern physics: the relevant processes are known as photon pair production and electron-positron annihilation. It is conceivable that photons might couple similarly to other kinds of matter. Just such a coupling to particles known as axions has recently been proposed in order to account for the apparent dimming of distant supernovae (see Chap. 6). Interactions of this kind cannot affect the overall level of intensity of the background radiation, as MacMillan thought, but they could leave characteristic traces in the *spectrum* of that radiation. These traces may be of great interest, as we will show in Chaps. 6–8.

## 1.14   Cosmic expansion

The last major new resolution to Olbers' paradox appeared with the relativistic steady-state theory of H. Bondi, T. Gold and F. Hoyle in 1948. This theory, like MacMillan's, postulated what was effectively continuous matter creation. However, atoms now did not arise out of interconversion from light; rather, they sprang from the vacuum itself. This followed from

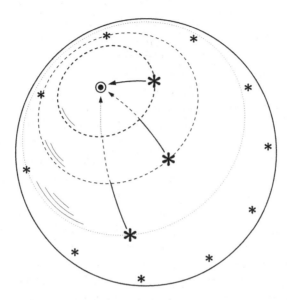

Fig. 1.14    Expansion. H. Bondi, T. Gold and F. Hoyle (1948) showed that in some cosmological models, the sky is dark at night because light from greater distances becomes progressively more weakened or *redshifted* by the expansion of the Universe.

the requirement that the Universe remain in a steady state as it expanded. (The theory therefore violated energy conservation continuously at low levels, but—as the theory's proponents were fond of pointing out—a similar phenomenon occurs in standard big-bang theory, only there it happens all at once.) Thus another explanation than MacMillan's had to be found for the darkness of the night sky. Bondi, Gold and Hoyle found such an explanation in the fact of *cosmic expansion*.

Expansion does two things to starlight from distant galaxies: it stretches its wavelength, and spaces photons more widely apart so that fewer of them reach us per unit time. Both effects reduce the intensity of starlight we see, and together they can significantly darken the night sky (Fig. 1.14). In the case of the steady-state theory, in fact, they are *entirely* responsible for making the sky dark at night. (This makes sense, since nothing can change in the steady-state Universe; there can be no phenomenon such as evolution in the number or luminosity of sources, or Olbers-type absorption and reheating, that would alter the brightness of the night sky with time.) Steady-state theorists therefore seized upon what they termed Olbers' paradox as the "basis of all modern cosmology" [29]. For them it truly was paradoxical; there was no way to understand the darkness of the night sky apart from cosmic expansion. Indeed, from the steady-state perspective, Olbers and all the others who pondered the dark night sky missed a great chance to *predict* that the Universe was expanding!

The steady-state theory, however, came into conflict with a growing body of observational evidence during the 1950s and was abandoned by most cosmologists after the discovery of the cosmic microwave background (CMB) in 1965. One might have expected the expansion resolution of Olbers' paradox to fade away at the same time. In fact, however, it proved far more durable than the steady-state theory itself. This may be because, as the first resolution to be proposed in the context of relativistic cosmology, it had somehow acquired an association with Einstein's theory of general relativity. Alternatively, the abstract notion of expanding space may have seemed more appropriately "modern" than alternatives based on the mundane physics of earlier centuries. But whatever the reason for its ongoing appeal, cosmic expansion is *not* the primary reason for the darkness of the night sky in the vast majority of realistic big-bang models, as was first forcefully argued by Harrison beginning in 1964 [26] and as we will make clear in Chaps. 2 and 3.

Two final points should be made in connection with expansion and the steady-state theory. First, while expansion is not primarily responsible for the darkness of the night sky as seen by human eyes, it does play a critical role in another part of the spectrum, namely the *microwave* band. Most of the light that reaches us from the depths of space in this band was not generated by stars over the lifetime of the Universe, but is instead believed to represent a "flash photo" of the big-bang fireball itself, cooled to microwave temperatures by billions of years of cosmic expansion. Thus it is expansion, not lifetime, that makes the microwave sky dark. Needless to say, Olbers himself was concerned with the visible sky and knew nothing of the CMB. However, it is in this sense that some authors have presented expansion as the resolution to a different kind of paradox in which, to quote S. Weinberg, "the microwave background appears as the pale image of the fiery furnace with which we were threatened by de Chéseaux and Olbers" [30, 31]. We will discuss the microwave background and possible contributions to it in Chap. 5.

The second point is that, whatever their original motivation, steady-state theorists performed a valuable service in resurrecting Olbers' argument and restoring it to the forefront of astronomy. There is now general agreement that any cosmology that hopes to be taken seriously must have a satisfactory explanation for the dark night sky. Indeed, as we will argue in the chapters to come, the converse holds as well: the search for unexplained features in the light of the night sky may hold the key to a deeper understanding of cosmology.

## 1.15    Olbers' paradox today

By 1970, only two of the dozen or more explanations originally proposed for the darkness of the night sky were still in active circulation: expansion and the finite age of the galaxies (or alternatively, the finite energy they have emitted). But widespread unease lingered as to which solution was correct, or more precisely, which solution was *more important* in determining the actual level of intensity of light reaching the Milky Way. There were debates over the fine points of Harrison's arguments [32–34]. It was pointed out that expansion could still be the dominant effect in some big-bang models [35], and that finite age could still be seen as the dominant effect in steady-state models—if "age" were defined in a technically more general way involving the concept of geodesic completeness [14].

Among scientific experts, discourse of this kind is a healthy sign of a discipline in its growing years. The net result for non-specialists and students of astronomy, however, was confusion. A survey of modern astronomy textbooks carried out in 1987 revealed that 30% agreed with Harrison in attributing the darkness of the night sky to the finite age of the galaxies, while 20% cited expansion as the cause. The other 50% mentioned both factors, though without an assessment of which was more important. This confusion prompted the appearance in 1987 of an article by P.S. Wesson, K. Valle and R. Stabell [36] that introduced a new method, to be presented in Chaps. 2 and 3, of tackling the problem of the dark night sky. With this method, the authors gave a simple yet exact quantitative assessment of the relative importance of the age and expansion factors, and thereby laid Olbers' paradox to rest with appropriate honors [37, 38].

We now know what Olbers did not: that the main reason why the sky is dark at night is that the Universe had a *beginning in time*. This can be appreciated qualitatively (and quantitatively to within a factor of a few) with no relativity at all beyond the fact of a finite speed of light. Imagine yourself at the center of a ball of radius $R$, filled with bright sources whose uniform luminosity density $\mathcal{L}(r) = \mathcal{L}_0$. The intensity $Q$ of background radiation between you and the edge of the ball is just

$$Q = \int_0^R \mathcal{L}(r)dr = ct_0\mathcal{L}_0 \,, \tag{1.1}$$

where we have used $R = ct_0$ as a naive approximation to the size of the Universe. Thus knowledge of the luminosity density $\mathcal{L}_0$ and measurement of the background intensity $Q$ tells us immediately that the galaxies have been shining only for a time $t_0$.

More refined calculations introduce only minor changes to this result. Expansion stretches the path length $R$, but this is more than offset by the dilution of the luminosity density $\mathcal{L}(r)$, which drops by roughly the same factor cubed. There is a further reduction in $\mathcal{L}(r)$ due to the redshifting of light from distant sources. So Eq. (1.1) represents a theoretical upper limit on the background intensity. In a fully general-relativistic treatment, one obtains the following expression for $Q$ in standard cosmological models whose scale factor varies as a power-law function of time ($R \propto t^\ell$):

$$Q = \frac{ct_0\mathcal{L}_0}{1+\ell} \,. \tag{1.2}$$

This result may be checked using Eq. (2.10) in Chap. 2. Thus Eq. (1.1) overestimates $Q$ as a function of $t_0$ by a factor of 5/3 in a Universe filled with dust-like matter ($\ell = 2/3$).

Insofar as $Q$ and $\mathcal{L}_0$ are both known quantities, one can in principle use them to infer a value for $t_0$. Intensity $Q$, for instance, is obtained by measuring spectral intensity $I_\lambda(\lambda)$ over the wavelengths where starlight is brightest and integrating: $Q = \int I_\lambda(\lambda)d\lambda$. This typically leads to values of around $Q \approx 1.4 \times 10^{-4}$ erg s$^{-1}$ cm$^{-2}$ [39]. Luminosity density $\mathcal{L}_0$ can be determined by counting the number of faint galaxies in the sky down to some limiting magnitude, and extrapolating to fainter magnitudes based on assumptions about the true distribution of galaxy luminosities. One finds in this way that $\mathcal{L}_0 \approx 1.9 \times 10^{-32}$ erg s$^{-1}$ cm$^{-3}$ [40]. Alternatively, one can extrapolate from the properties of the Sun, which emits its energy at a rate per unit mass of $\epsilon_\odot = L_\odot/M_\odot = 1.9$ erg s$^{-1}$ g$^{-1}$. A color-magnitude diagram for nearby stars shows us that the Sun is modestly brighter than average, so a more typical rate of stellar energy emission is about 1/4 the Solar value or $\epsilon \sim 0.5$ erg s$^{-1}$ g$^{-1}$. Multipying this number by the density of luminous matter in the Universe ($\rho_{\rm lum} = 4 \times 10^{-32}$ g cm$^{-3}$) gives a figure for mean luminosity density which is the same as that derived above from galaxy counts: $\mathcal{L}_0 = \epsilon\rho_{\rm lum} \sim 2 \times 10^{-32}$ erg s$^{-1}$ cm$^{-3}$. Either way, plugging $Q$ and $\mathcal{L}_0$ into Eq. (1.1) with $\ell = 2/3$ implies a cosmic age of $t_0 = 13$ Gyr,

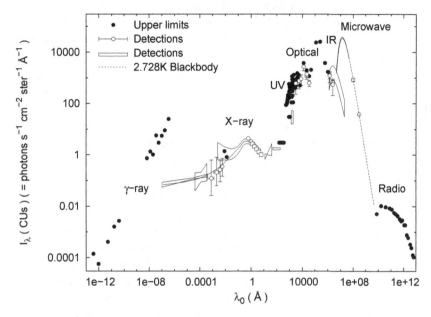

Fig. 1.15   A compilation of experimental measurements of the intensity of cosmic background radiation at all wavelengths. This figure and the data shown in it will be discussed in more detail in later sections.

which differs from the currently accepted figure by only 5%. (The remaining difference can be accounted for if cosmic expansion is not a simple power-law function of time; more on this later.) *Thus the brightness of the night sky tells us not only that there was a big bang, but also roughly when it occurred.* Conversely, the intensity of background radiation is largely determined by the age of the Universe. Expansion merely deepens the shade of a night sky that is already dark.

We have so far discussed only the bolometric, or integrated intensity of the background light over all wavelengths, whose significance will be explored in more detail in Chap. 2. The *spectral* background—from radio to microwave, infrared, optical, ultraviolet, x-ray and gamma-ray bands— represents an even richer store of information about the Universe and its contents (Fig. 1.15). The optical waveband (where galaxies emit most of their light) has been of particular importance historically, and the infrared band (where the redshifted light of distant galaxies actually reaches us) has come into prominence recently. By combining the observational data in both of these bands, we can piece together much of the evolutionary history of the galaxy population, make inferences about the nature of the intervening intergalactic medium, and draw conclusions about the dynamical history of the Universe itself. Interest in this subject has exploded over the past few years as improvements in telescope and detector technology have brought us to the threshold of the first EBL detection in the optical and infrared bands. These developments and their implications are discussed in Chap. 3.

In the remainder of our account, we move on to what the background radiation tells us about the dark matter and energy, whose current status is reviewed in Chap. 4. The leading candidates are taken up individually in Chaps. 5–9. *None of them are perfectly black.* All of them are capable in principle of decaying into or interacting with ordinary photons, thereby leaving telltale signatures in the spectrum of background radiation. We begin with dark energy, for which there is particularly good reason to suspect a decay with time. The most likely place to look for evidence of such a process is in the cosmic microwave background, and we review the stringent constraints that can be placed on any such scenario in Chap. 5. Axions, neutrinos and weakly interacting massive particles are treated next: these particles decay into photons in ways that depend on parameters such as coupling strength, decay lifetime, and rest mass. As we show in Chaps. 6, 7 and 8, data in the infrared, optical, ultraviolet, x-ray and gamma-ray bands allow us to be specific about the kinds of properties that these particles must have if they are to make up the dark matter in the Universe. In Chap. 9, finally, we turn to black holes. The observed intensity of background radiation, especially in the gamma-ray band, is sufficient to rule out a significant role for standard four-dimensional black holes; but it may be possible for their higher-dimensional analogs (known as solitons) to make up all or part of the dark matter. We conclude in Chap. 10 with some final comments and a view toward future developments.

Chapter 2

# The Intensity of Cosmic Background Light

## 2.1 Bolometric intensity

Let us begin with the general problem of adding up the contributions from many sources of radiation in the Universe in such a way as to arrive at their combined intensity as received by us in the Milky Way (Fig. 2.1). To begin with, we take the sources to be ordinary galaxies, but the formalism is general. Consider a single galaxy at coordinate distance $r$ whose luminosity, or rate of energy emission per unit time, is given by $L(t)$. In a standard Friedmann-Lemaître-Roberston-Walker (FLRW) Universe, its energy has been spread over an area

$$A = \int dA = \int_{\theta=0}^{\pi} \int_{\phi=0}^{2\pi} [R_0 r d\theta][R_0 r \sin\theta d\phi] = 4\pi R_0^2 r^2 \qquad (2.1)$$

by the time it reaches us at $t = t_0$. Here we follow standard practice and use the subscript "0" to denote any quantity taken at the present time, so $R_0 \equiv R(t_0)$ is the present value of the cosmological scale factor $R$.

The intensity, or flux of energy per unit area reaching us from this galaxy is given by

$$dQ_g = \left[\frac{R(t)}{R_0}\right]^2 \frac{L(t)}{A} = \frac{R^2(t)L(t)}{4\pi R_0^4 r^2} \cdot \qquad (2.2)$$

Here the subscript "$g$" denotes a single galaxy, and the two factors of $R(t)/R_0$ reflect the fact that expansion increases the wavelength of the light as it travels toward us (reducing its energy), and also spaces the photons more widely apart in time (Hubble's energy and number effects).

To describe the whole population of galaxies, distributed through space with physical number density $n_g(t)$, it is convenient to use the four-dimensional galaxy current $J_g^\mu \equiv n_g U^\mu$ where $U^\mu \equiv (1,0,0,0)$ is the galaxy

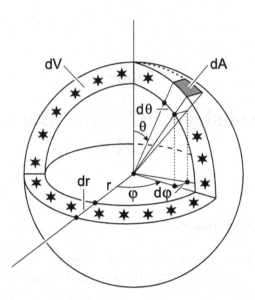

Fig. 2.1   Surface area element $dA$ and volume $dV$ of a thin spherical shell, containing all the light sources at coordinate distance $r$.

four-velocity [30]. If galaxies are conserved (i.e. their rates of formation and destruction by merging or other processes are slow in comparison to the expansion rate of the Universe), then $J_g^\mu$ obeys a conservation equation

$$\nabla_\mu J_g^\mu = 0 \,, \tag{2.3}$$

where $\nabla_\mu$ denotes the covariant derivative. Using the Robertson-Walker metric this reduces to:

$$\frac{1}{R^3}\frac{d}{dt}\left(R^3 n_g\right) = 0 \,. \tag{2.4}$$

In what follows, we will replace $n_g$ by the *comoving number density* $(n)$, defined in terms of $n_g$ by

$$n \equiv n_g \left(\frac{R}{R_0}\right)^3 \,. \tag{2.5}$$

Under the assumption of galaxy conservation, this quantity is always equal to its present value ($n = n_0 = $ const). When merging or other galaxy non-conserving processes are important, $n$ is no longer constant. We will allow for this situation in Sec. 3.7.

We can take ourselves as located at the center of the spherical shell in Fig. 2.1, and consider those galaxies located inside the shell which extends from radial coordinate distance $r$ to $r + dr$. The volume of this shell is obtained from the Robertson-Walker metric as

$$dV = \int_{\theta=0}^{\pi} \int_{\phi=0}^{2\pi} \left[ \frac{Rdr}{\sqrt{1-kr^2}} \right] [Rrd\theta] [Rr\sin\theta d\phi] = \frac{4\pi R^3 r^2 dr}{\sqrt{1-kr^2}} . \qquad (2.6)$$

The only trajectories of interest are those of light rays striking our detectors at the origin. By definition, these are radial ($d\theta = d\phi = 0$) null geodesics ($ds^2 = 0$), for which the metric relates time $t$ and coordinate distance $r$ via

$$cdt = \frac{R\,dr}{\sqrt{1-kr^2}} . \qquad (2.7)$$

Thus the volume of the shell can be re-expressed as

$$dV = 4\pi R^2 r^2 cdt , \qquad (2.8)$$

and the latter may now be thought of as extending between look-back times $t_0 - t$ and $t_0 - (t + dt)$, rather than distances $r$ and $r + dr$.

The total energy received at the origin from the galaxies in the shell is then just the product of their individual intensities (2.2), their number per unit volume (2.5), and the volume of the shell (2.8):

$$dQ = dQ_g\, n_g\, dV = cn(t)\tilde{R}(t)L(t)dt . \qquad (2.9)$$

Here we have defined the normalized scale factor by $\tilde{R} \equiv R/R_0$. We henceforth use tildes throughout this book to denote dimensionless quantities *taken relative to their present value* at $t_0$.

Integrating (2.9) over all shells between $t_0$ and $t_0 - t_f$, where $t_f$ is the source-formation time, we obtain

$$Q = \int dQ = c \int_{t_f}^{t_0} n(t)\, L(t)\, \tilde{R}(t)\, dt . \qquad (2.10)$$

This defines the *bolometric intensity of the extragalactic background light* (EBL). This is the energy received by us (over all wavelengths of the electromagnetic spectrum) per unit time, per unit area, from all the galaxies which have been shining since time $t_f$. In principle, if we let $t_f \to 0$, we will encompass the entire history of the Universe since the big bang. Although this sometimes provides a useful mathematical shortcut, we will see in later sections that it is physically more realistic to cut the integral off at a finite formation time. The quantity $Q$ is a measure of the amount of light in the Universe, and Olbers' "paradox" is merely another way of asking why it is low.

## 2.2    Time and redshift

While the cosmic time $t$ is a useful independent variable for theoretical purposes, it is not directly observable. In studies aimed at making contact with eventual observation it is better to work in terms of redshift $z$, which is the relative shift in wavelength $\lambda$ of a light signal between the time it is emitted and observed:

$$z \equiv \frac{\Delta\lambda}{\lambda} = \frac{R_0 - R(t)}{R(t)} = \tilde{R}^{-1} - 1 \,. \qquad (2.11)$$

Differentiating with resepct to time, and defining the Hubble expansion rate $H \equiv \dot{R}/R$, we find that

$$dt = -\frac{dz}{(1+z)H(z)} \,. \qquad (2.12)$$

Hence Eq. (2.10) is converted into an integral over $z$ as

$$Q = c \int_0^{z_f} \frac{n(z)\,L(z)\,dz}{(1+z)^2\,H(z)} \,, \qquad (2.13)$$

where $z_f$ is the redshift of galaxy formation.

For some problems, and for this section in particular, the physics of the sources themselves are of secondary importance, and it is reasonable to take $L(z) = L_0$ and $n(z) = n_0$ as constants over the range of redshifts of interest. Then Eq. (2.13) can be written in the form

$$Q = Q_* \int_0^{z_f} \frac{dz}{(1+z)^2\,\tilde{H}(z)} \,. \qquad (2.14)$$

Here $\tilde{H} \equiv H/H_0$ is the Hubble parameter, normalized to its present value, and $Q_*$ is a constant containing all the dimensional information:

$$Q_* \equiv \frac{c\mathcal{L}_0}{H_0} \,. \qquad (2.15)$$

The quantities $H_0$, $\mathcal{L}_0$ and $Q_*$ are fundamental to much of what follows, and we pause briefly to discuss them here. The value of $H_0$ (Hubble's "constant") is still debated, and is commonly expressed in the form

$$H_0 = 100\,h_0 \text{ km s}^{-1}\text{Mpc}^{-1} = 0.102\,h_0 \text{ Gyr}^{-1} \,. \qquad (2.16)$$

Here $1 \text{ Gyr} \equiv 10^9$ yr and the uncertainties have been absorbed into a dimensionless parameter $h_0$ whose value is now conservatively estimated to lie in the range $0.6 \leqslant h_0 \leqslant 0.9$ (Sec. 4.2).

The quantity $\mathcal{L}_0$ is the *comoving luminosity density* of the Universe at $z = 0$:

$$\mathcal{L}_0 \equiv n_0 L_0 . \tag{2.17}$$

This can be measured experimentally by counting galaxies down to some limiting apparent magnitude, and extrapolating to fainter ones based on assumptions about the true distribution of absolute magnitudes. A compilation of seven such studies gives [40]:

$$\mathcal{L}_0 = (2.0 \pm 0.2) \times 10^8 \, h_0 \, L_\odot \, \mathrm{Mpc}^{-3}$$

$$= (2.6 \pm 0.3) \times 10^{-32} \, h_0 \, \mathrm{erg \, s}^{-1} \, \mathrm{cm}^{-3} , \tag{2.18}$$

near 4400Å in the B-band, where galaxies emit most of their light. We will use this number throughout this book. Measurements for the Two Degree Field (2dF) suggest a slightly lower value, $\mathcal{L}_0 = (1.82 \pm 0.17) \times 10^8 \, h_0 \, L_\odot \, \mathrm{Mpc}^{-3}$ [41], and this agrees with figures from the Sloan Digital Sky Survey (SDSS): $\mathcal{L}_0 = (1.84 \pm 0.04) \times 10^8 \, h_0 \, L_\odot \, \mathrm{Mpc}^{-3}$ [42]. If the final result inferred from large-scale galaxy surveys of this kind proves to be significantly different from that in (2.18), then our EBL intensities (which are proportional to $\mathcal{L}_0$) would go up or down accordingly.

Using (2.16) and (2.18), we find that the characteristic intensity associated with the integral (2.14) takes the value

$$Q_* \equiv \frac{c\mathcal{L}_0}{H_0} = (2.5 \pm 0.2) \times 10^{-4} \, \mathrm{erg \, s}^{-1} \, \mathrm{cm}^{-2} . \tag{2.19}$$

There are two important things to note about this quantity. First, because the factors of $h_0$ attached to both $\mathcal{L}_0$ and $H_0$ cancel each other out, $Q_*$ is *independent* of the uncertainty in Hubble's constant. This is not always appreciated but was first emphasized by J.E. Felten [43]. Second, the value of $Q_*$ is *very small* by everyday standards: more than a million times fainter than the bolometric intensity that would be produced by a 100 W bulb in the middle of an ordinary-size living room whose walls, floor and ceiling have a summed surface area of 100 m$^2$ ($Q_{\mathrm{bulb}} = 10^3 \, \mathrm{erg \, s}^{-1} \, \mathrm{cm}^{-2}$). The smallness of $Q_*$ is intimately related to the resolution of Olbers' paradox.

## 2.3 Matter, energy and expansion

The remaining unknown in Eq. (2.14) is the relative expansion rate $\tilde{H}(z)$, which is obtained by solving the field equations of general relativity. For standard FLRW models one obtains the following differential equation:

$$\tilde{H}^2 + \frac{kc^2}{H_0^2 R^2} = \frac{8\pi G}{3H_0^2} \rho_{\mathrm{tot}} . \tag{2.20}$$

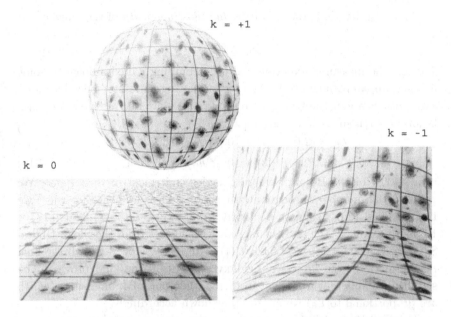

Fig. 2.2   The 3D spatial curvature of the Universe (here represented by 2D surfaces) is determined by its total density, expressed in units of the critical density $\rho_{\mathrm{crit},0}$ by the parameter $\Omega_{\mathrm{tot},0}$. If $\Omega_{\mathrm{tot},0} = 1$ then 3D space is flat ("Euclidean") and the curvature parameter $k = 0$. If $\Omega_{\mathrm{tot},0} > 1$ then the 3D Universe is positively curved (spherical) and $k = +1$. If $\Omega_{\mathrm{tot},0} < 1$ then it is negatively curved (hyperbolic) and $k = -1$. (Figure courtesy E.C. Eekels.)

Here $\rho_{\mathrm{tot}}$ is the total density of all forms of matter-energy, including the density of vacuum energy associated with the cosmological constant $\Lambda$ via

$$\rho_\Lambda c^2 \equiv \frac{\Lambda c^4}{8\pi G} = \text{const}.\qquad(2.21)$$

It is convenient to define the present *critical density* by

$$\rho_{\mathrm{crit},0} \equiv \frac{3H_0^2}{8\pi G} = (1.88 \times 10^{-29})\,h_0^2\ \mathrm{g\ cm^{-3}}.\qquad(2.22)$$

The value of $\rho_{\mathrm{crit},0}$ determines the spatial curvature of the Universe we live in (Fig. 2.2). If the densities of all forms of matter-energy sum to $\rho_{\mathrm{crit},0}$, then $k = 0$ and the ordinary three-dimensional (3D) space of the Universe is flat or Euclidean. Such a Universe is unbounded and infinite in extent. Alternatively, if the total density exceeds $\rho_{\mathrm{crit},0}$, then $k > 0$ and the 3D Universe is positively curved. This is usually referred to as a "$k = +1$ model" because the magnitude of the curvature constant $k$ can be normalized to unity by a choice of units for $R_0$. Spatial hypersurfaces in this model are spherical in shape, and the Universe is closed and finite but unbounded. Finally, if all forms of matter-energy have a combined density less than $\rho_{\text{crit},0}$, then the Universe has negative curvature ($k = -1$) and hyperbolic spatial sections. It is open, unbounded and infinite in extent.

Cosmological densities $\rho$ are commonly expressed in the form of dimensionless *density parameters*, $\Omega \equiv \rho/\rho_{\text{crit},0}$. The density parameter for all forms of matter-energy combined is $\Omega_{\text{tot}} = \rho_{\text{tot}}/\rho_{\text{crit},0}$, with a present value $\Omega_{\text{tot},0}$. Using the latter parameter and evaluating Eq. (2.20) at the present time, we can eliminate the unknown $k$ via $kc^2/(H_0 R_0)^2 = \Omega_{\text{tot},0} - 1$. Substituting this result back into (2.20) produces:

$$\tilde{H}^2 = \Omega_{\text{tot}} - (\Omega_{\text{tot},0} - 1)\,\tilde{R}^{-2}\,. \tag{2.23}$$

To complete the problem we need only the form of $\Omega_{\text{tot}}(\tilde{R})$, which comes from energy conservation. Under the usual assumptions of isotropy and homogeneity, the matter-energy content of the Universe can be modelled by an energy-momentum tensor of the *perfect-fluid* form:

$$T_{\mu\nu} = (\rho + p/c^2)U_\mu U_\nu + p\,g_{\mu\nu}\,. \tag{2.24}$$

Here the density $\rho$ and pressure $p$ are related by an equation of state, which is commonly written as

$$p = (\gamma - 1)\rho c^2\,. \tag{2.25}$$

Three equations of state are of particular relevance to cosmology, and will make regular appearances in the sections that follow:

$$\gamma = \begin{cases} 4/3 & \Rightarrow & p_r = \rho_r c^2/3 & \text{(radiation)} \\ 1 & \Rightarrow & p_m = 0 & \text{(dust-like matter)} \\ 0 & \Rightarrow & p_v = -\rho_v\,c^2 & \text{(vacuum energy)} \end{cases} \tag{2.26}$$

The first of these is a good approximation to the *early* Universe, when conditions were so hot and dense that matter and radiation existed in nearly perfect thermodynamic equilibrium (the radiation era). The second has often been taken to describe the *present* Universe, since we know that the energy density of electromagnetic radiation now is far below that of dust-like matter. The third may be a good description of the *future* state of the Universe, if recent measurements of the magnitudes of high-redshift Type Ia supernovae are borne out (Chap. 4). These indicate that vacuum-like dark energy is already more important than all other contributions to the density of the Universe combined.

Assuming that energy and momentum are neither created nor destroyed, one can proceed exactly as with the galaxy current $J_g^\mu$. The conservation equation now reads

$$\nabla^\mu T_{\mu\nu} = 0 \ . \tag{2.27}$$

With the definition (2.24) this reduces to

$$\frac{1}{R^3} \frac{d}{dt} \left[ R^3 \left( \rho c^2 + p \right) \right] = \frac{dp}{dt} \ , \tag{2.28}$$

which may be compared with (2.4) for galaxies. Eq. (2.28) is solved with the help of the equation of state (2.25) to yield

$$\rho = \rho_0 \tilde{R}^{-3\gamma} \ . \tag{2.29}$$

In particular, for the single-component fluids in (2.26):

$$\begin{aligned}
\rho_r &= \rho_{r,0} \tilde{R}^{-4} && \text{(radiation)} \\
\rho_m &= \rho_{m,0} \tilde{R}^{-3} && \text{(dust-like matter)} \ . \\
\rho_v &= \rho_{v,0} = \text{const} && \text{(vacuum energy)}
\end{aligned} \tag{2.30}$$

These expressions will frequently prove useful in later sections. They are also applicable to cases in which several components are present, as long as these components exchange energy slowly relative to the expansion rate, so that each is in effect conserved separately.

The expansion rate (2.23) can be expressed in terms of redshift $z$ with the help of Eqs. (2.11) and (2.30) as follows:

$$\tilde{H}(z) = \left[ \Omega_{r,0}(1 + z)^4 + \Omega_{m,0}(1 + z)^3 + \Omega_{\Lambda,0} \right.$$
$$\left. - \left(\Omega_{\text{tot},0} - 1\right)(1 + z)^2 \right]^{1/2} \ . \tag{2.31}$$

Here $\Omega_{\Lambda,0} \equiv \rho_\Lambda/\rho_{\text{crit},0} = \Lambda c^2/3H_0^2$ from (2.21) and (2.22). Eq. (2.31) is sometimes referred to as the *Friedmann-Lemaître equation*.

The radiation ($\Omega_{r,0}$) term can be neglected in all but earliest stages of cosmic history ($z \gtrsim 100$), since $\Omega_{r,0}$ is some four orders of magnitude smaller than $\Omega_{m,0}$. The vacuum ($\Omega_{\Lambda,0}$) term in Eq. (2.31) is independent of redshift, which means that its influence is not diluted with time. Any Universe with $\Lambda > 0$ will therefore eventually be dominated by vacuum energy. In the limit $t \to \infty$, in fact, the Friedmann-Lemaître equation. reduces to $\Omega_{\Lambda,0} = (H_\infty/H_0)^2$, where $H_\infty$ is the limiting value of $H(t)$ as $t \to \infty$ (assuming that this latter limit exists; i.e. that the Universe does not recollapse). It follows that

$$\Lambda c^2 = 3H_\infty^2 \ . \tag{2.32}$$

If $\Lambda > 0$, then we will *necessarily* measure $\Omega_{\Lambda,0} \sim 1$ at late times, regardless of the microphysical origin of the vacuum energy.

It was common during the 1980s to work with a simplified version of Eq. (2.31), in which not only the radiation term was neglected, but the vacuum ($\Omega_{\Lambda,0}$) and curvature ($\Omega_{\text{tot},0}$) terms as well. There were four principal reasons for the popularity of this *Einstein-de Sitter* (EdS) model. First, all four terms on the right-hand side of Eq. (2.31) depend differently on $z$, so it would seem surprising to find ourselves living in an era when any two of them were of comparable size. By this argument, which goes back to R.H. Dicke [44], it was felt that one term ought to dominate at any given time. Second, the vacuum term was regarded with particular suspicion for reasons to be discussed in Sec. 4.5. Third, a period of inflation was asserted to have driven $\Omega_{\text{tot}}(t)$ to unity. (This is still widely believed, but depends on the initial conditions preceding inflation, and does not necessarily hold in all plausible models [45, 46].) And finally, the EdS model was favored on grounds of simplicity. These arguments are no longer compelling today, and the determination of $\Omega_{m,0}$ and $\Omega_{\Lambda,0}$ has shifted largely back into the empirical domain. We discuss the observational status of these constants in more detail in Chap. 4, merely assuming here that radiation and matter densities are positive and not too large ($0 \leqslant \Omega_{r,0} \leqslant 1.5$ and $0 \leqslant \Omega_{m,0} \leqslant 1.5$), and that vacuum energy density is neither too large nor too negative ($-0.5 \leqslant \Omega_{\Lambda,0} \leqslant 1.5$).

## 2.4  How important is expansion?

Eq. (2.14) provides us with a simple integral for the bolometric intensity $Q$ of the extragalactic background light in terms of the constant $Q_*$, Eq. (2.19), and the expansion rate $\tilde{H}(z)$, Eq. (2.31). On dimensional grounds, we would expect $Q$ to be close to $Q_*$ as long as the function $\tilde{H}(z)$ is sufficiently well-behaved over the lifetime of the galaxies, and we will find that this expectation is borne out in all realistic cosmological models.

"Olbers' paradox" is essentially the question of why $Q$ is so small (of order $Q_*$). In the context of modern big-bang cosmology, as discussed in Chap. 1, the answer can boil down to only one of two things: the *finite age* of the Universe (which limits the total amount of light that has been produced) or *cosmic expansion* (which dilutes the intensity of intergalactic radiation, and also redshifts the light signals from distant sources).

The relative importance of the two factors has continued to be a subject of controversy [36]. In particular there is a lingering perception that general relativity "solves" Olbers' paradox chiefly because the expansion of the Universe stretches and dims the light it contains.

There is a simple way to test this supposition using the formalism we have already laid out, and that is to "turn off" the expansion by setting the scale factor of the Universe equal to a constant value, $R(t) = R_0$. Then $\tilde{R} = 1$ and Eq. (2.10) gives the bolometric intensity of the extragalactic background light (EBL) as

$$Q_{\text{stat}} = Q_* \, H_0 \int_{t_f}^{t_0} dt \, . \tag{2.33}$$

Here we have taken $n = n_0$ and $L = L_0$ as before, and used (2.19) for $Q_*$. The subscript "stat" denotes the *static analog* of $Q$; that is, the intensity that one would measure in a Universe which did not expand. Eq. (2.33) shows that this is just the length of time for which the galaxies have been shining, measured in units of Hubble time $(H_0^{-1})$ and scaled by $Q_*$.

We wish to compare (2.14) in the expanding Universe with its static analog (2.33), while keeping all other factors the same. In particular, if the comparison is to be meaningful, the *lifetime* of the galaxies should be identical. This is just $\int dt$, which may—in an expanding Universe—be converted to an integral over redshift $z$ by means of (2.12):

$$\int_{t_f}^{t_0} dt = \frac{1}{H_0} \int_0^{z_f} \frac{dz}{(1+z)\tilde{H}(z)} \, . \tag{2.34}$$

In a static Universe, of course, redshift does not carry its usual physical significance. But nothing prevents us from retaining $z$ as an integration variable. Substitution of (2.34) into (2.33) then yields

$$Q_{\text{stat}} = Q_* \int_0^{z_f} \frac{dz}{(1+z)\tilde{H}(z)} \, . \tag{2.35}$$

We emphasize that $z$ and $\tilde{H}$ are to be seen here as algebraic parameters whose usefulness lies in the fact that they ensure consistency in *age* between the static and expanding pictures.

Eq. (2.14) and its static analog (2.35) allow us to isolate the relative importance of expansion versus lifetime, by evaluating the *ratio* $Q/Q_{\text{stat}}$ for all reasonable values of the cosmological parameters $\Omega_{r,0}, \Omega_{m,0}$ and $\Omega_{\Lambda,0}$. If $Q/Q_{\text{stat}} \ll 1$ over most of this phase space, then expansion must reduce $Q$ significantly from what it would otherwise be in a static Universe. Conversely, values of $Q/Q_{\text{stat}} \approx 1$ would tell us that expansion has little effect, and that (as in the static case) the brightness of the night sky is determined primarily by the length of time for which the galaxies have been shining.

| Model Name | $\Omega_{r,0}$ | $\Omega_{m,0}$ | $\Omega_{\Lambda,0}$ | $1 - \Omega_{tot,0}$ |
|---|---|---|---|---|
| Radiation | 1 | 0 | 0 | 0 |
| Einstein-de Sitter | 0 | 1 | 0 | 0 |
| de Sitter | 0 | 0 | 1 | 0 |
| Milne | 0 | 0 | 0 | 1 |

## 2.5   Simple flat models

We begin by evaluating $Q/Q_{\text{stat}}$ for the simplest cosmological models, those in which the Universe has one critical-density component or contains nothing at all (Table 2.1). Consider first the *radiation model* with a critical density of radiation or ultra-relativistic particles ($\Omega_{r,0} = 1$) but $\Omega_{m,0} = \Omega_{\Lambda,0} = 0$. Bolometric EBL intensity in the expanding Universe is, from (2.14)

$$\frac{Q}{Q_*} = \int_1^{1+z_f} \frac{dx}{x^4} = \begin{cases} 21/64 & (z_f = 3) \\ 1/3 & (z_f = \infty) \end{cases} , \qquad (2.36)$$

where $x \equiv 1 + z$. The corresponding result for a static model is given by (2.35) as

$$\frac{Q_{\text{stat}}}{Q_*} = \int_1^{1+z_f} \frac{dx}{x^3} = \begin{cases} 15/32 & (z_f = 3) \\ 1/2 & (z_f = \infty) \end{cases} . \qquad (2.37)$$

Here we have chosen illustrative lower and upper limits for the redshift of galaxy formation ($z_f = 3$ and $\infty$ respectively). The actual value of this parameter has not yet been determined, although there are now indications that $z_f$ may be as high as six. In any case, it may be seen that overall EBL intensity is rather insensitive to this parameter. Increasing $z_f$ lengthens the period over which galaxies radiate, and this increases both $Q$ and $Q_{\text{stat}}$. The ratio $Q/Q_{\text{stat}}$, however, is given by

$$\frac{Q}{Q_{\text{stat}}} = \begin{cases} 7/10 & (z_f = 3) \\ 2/3 & (z_f = \infty) \end{cases} , \qquad (2.38)$$

and this changes but little. We will find this to be true in general.

Consider next the Einstein-de Sitter model, which has a critical density of dust-like matter ($\Omega_{m,0} = 1$) with $\Omega_{r,0} = \Omega_{\Lambda,0} = 0$. Bolometric EBL intensity in the expanding Universe is, from (2.14)

$$\frac{Q}{Q_*} = \int_1^{1+z_f} \frac{dx}{x^{7/2}} = \begin{cases} 31/80 & (z_f = 3) \\ 2/5 & (z_f = \infty) \end{cases} . \qquad (2.39)$$

The corresponding static result is given by (2.35) as

$$\frac{Q_{\text{stat}}}{Q_*} = \int_1^{1+z_f} \frac{dx}{x^{5/2}} = \begin{cases} 7/12 \ (z_f = 3) \\ 2/3 \ (z_f = \infty) \end{cases} . \qquad (2.40)$$

The ratio of EBL intensity in an expanding Einstein-de Sitter model to that in the equivalent static model is thus

$$\frac{Q}{Q_{\text{stat}}} = \begin{cases} 93/140 \ (z_f = 3) \\ 3/5 \ (z_f = \infty) \end{cases} . \qquad (2.41)$$

A third simple case is the *de Sitter model*, which consists entirely of vacuum energy ($\Omega_{\Lambda,0} = 1$), with $\Omega_{r,0} = \Omega_{m,0} = 0$. Bolometric EBL intensity in the expanding case is, from (2.14)

$$\frac{Q}{Q_*} = \int_1^{1+z_f} \frac{dx}{x^2} = \begin{cases} 3/4 \ (z_f = 3) \\ 1 \ (z_f = \infty) \end{cases} . \qquad (2.42)$$

Eq. (2.35) gives for the equivalent static case

$$\frac{Q_{\text{stat}}}{Q_*} = \int_1^{1+z_f} \frac{dx}{x} = \begin{cases} \ln 4 \ (z_f = 3) \\ \infty \ (z_f = \infty) \end{cases} . \qquad (2.43)$$

The ratio of EBL intensity in an expanding de Sitter model to that in the equivalent static model is then

$$\frac{Q}{Q_{\text{stat}}} = \begin{cases} 3/(4\ln 4) \ (z_f = 3) \\ 0 \ (z_f = \infty) \end{cases} . \qquad (2.44)$$

The de Sitter Universe is older than other models, which means it has more time to fill up with light, so intensities are higher. In fact, $Q_{\text{stat}}$ (which is proportional to the lifetime of the galaxies) goes to infinity as $z_f \to \infty$, driving $Q/Q_{\text{stat}}$ to zero in this limit. (It is thus possible in principle to "recover Olbers' paradox" in the de Sitter model [47].) Such a limit is however unphysical in the context of the EBL, since the ages of galaxies and their component stars are bounded from above. For realistic values of $z_f$ one obtains values of $Q/Q_{\text{stat}}$ which are only slightly lower than those in the radiation and matter cases.

Finally, we consider the *Milne model*, which is empty of all forms of matter and energy ($\Omega_{r,0} = \Omega_{m,0} = \Omega_{\Lambda,0} = 0$), making it an idealization but one which has often proved useful. Bolometric EBL intensity in the expanding case is given by (2.14) and turns out to be identical to Eq. (2.37) for the static radiation model. The corresponding static result, as given by (2.35), turns out to be the same as Eq. (2.42) for the expanding de Sitter model. The ratio of EBL intensity in an expanding Milne model to that in the equivalent static model is then

$$\frac{Q}{Q_{\text{stat}}} = \begin{cases} 5/8 \ (z_f = 3) \\ 1/2 \ (z_f = \infty) \end{cases} . \tag{2.45}$$

This again lies close to previous results. In all cases (except the $z_f \to \infty$ limit of the de Sitter model) the ratio of bolometric EBL intensities with and without expansion lies in the range $0.4 \lesssim Q/Q_{\text{stat}} \lesssim 0.7$. This shows that expansion is not an important influence in the intensity of background light, at least in simple models.

## 2.6 Curved and multi-fluid models

To see whether the pattern observed in the previous section holds more generally, we expand our investigation to the wider class of open and closed models. Eqs. (2.14) and (2.35) may be solved analytically for these cases, if they are dominated by a single component [39]. We plot the results in Figs. 2.3, 2.4 and 2.5 for radiation-, matter- and vacuum-dominated models respectively. In each figure, long-dashed lines correspond to EBL intensity in expanding models $(Q/Q_*)$, while short-dashed ones show the equivalent static quantities $(Q_{\text{stat}}/Q_*)$. The ratio of these two quantities $(Q/Q_{\text{stat}})$ is indicated by solid lines. Heavy lines have $z_f = 3$ while light ones are calculated for $z_f = \infty$.

Figs. 2.3 and 2.4 show that while the individual intensities $Q/Q_*$ and $Q_{\text{stat}}/Q_*$ do vary significantly with $\Omega_{r,0}$ and $\Omega_{m,0}$ in radiation- and matter-dominated models, their ratio $Q/Q_{\text{stat}}$ remains nearly constant across the whole of the phase space, staying inside the range $0.5 \lesssim Q/Q_{\text{stat}} \lesssim 0.7$ for both models.

Fig. 2.5 shows that a similar trend occurs in vacuum-dominated models. While absolute EBL intensities $Q/Q_*$ and $Q_{\text{stat}}/Q_*$ differ from those in the radiation- and matter-dominated models, their *ratio* (solid lines) is again close to flat. The exception occurs as $\Omega_{\Lambda,0} \to 1$ (de Sitter model), where $Q/Q_{\text{stat}}$ dips well below 0.5 for large $z_f$. For $\Omega_{\Lambda,0} > 1$, there is no big bang (in models with $\Omega_{r,0} = \Omega_{m,0} = 0$), and one has instead a "big bounce" (i.e. a nonzero scale factor at the beginning of the expansionary phase). This implies a maximum possible redshift $z_{\text{max}}$ given by

$$1 + z_{\text{max}} = \sqrt{\frac{\Omega_{\Lambda,0}}{\Omega_{\Lambda,0} - 1}} . \tag{2.46}$$

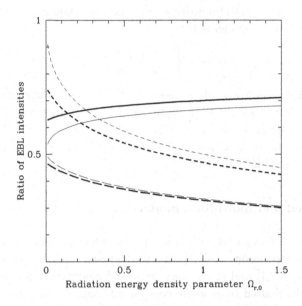

Fig. 2.3   Ratios $Q/Q_*$ (long-dashed lines), $Q_{\text{stat}}/Q_*$ (short-dashed lines) and $Q/Q_{\text{stat}}$
(solid lines) as a function of radiation density $\Omega_{r,0}$. Heavier lines are calculated for
$z_f = 3$ while lighter ones have $z_f = \infty$.

While such models are rarely considered, it is interesting to note that the
same pattern persists here, albeit with one or two wrinkles. In view of
(2.46), one can no longer integrate out to arbitrarily high formation-redshift
$z_f$. If one wants to integrate to *at least* $z_f$, then one is limited to vacuum
densities less than $\Omega_{\Lambda,0} < (1 + z_f)^2/[(1 + z_f)^2 - 1]$, or $\Omega_{\Lambda,0} < 16/15$ for the
case $z_f = 3$ (heavy dotted line). More generally, for $\Omega_{\Lambda,0} > 1$ the limiting
value of EBL intensity (shown with light lines) is reached as $z_f \to z_{\text{max}}$
rather than $z_f \to \infty$ for both expanding and static models. Over the entire
parameter space $-0.5 \leqslant \Omega_{\Lambda,0} \leqslant 1.5$ (except in the immediate vicinity of
$\Omega_{\Lambda,0} = 1$), Fig. 2.5 shows that $0.4 \lesssim Q/Q_{\text{stat}} \lesssim 0.7$ as before.

When more than one component of matter is present, analytic expressions can be found in only a few special cases, and the ratios $Q/Q_*$ and
$Q_{\text{stat}}/Q_*$ must in general be computed numerically. We show the results
in Fig. 2.6 for the situation which is of most physical interest: a Universe
containing both dust-like matter ($\Omega_{m,0}$, horizontal axis) and vacuum energy ($\Omega_{\Lambda,0}$, vertical axis), with $\Omega_{r,0} = 0$. This is a contour plot, with five
bundles of equal-EBL intensity contours for the expanding Universe (la-
belled $Q/Q_* = 0.37, 0.45, 0.53, 0.61$ and $0.69$). The heaviest (solid) lines

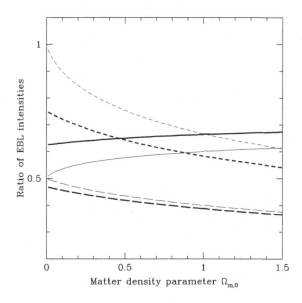

Fig. 2.4 Ratios $Q/Q_*$ (long-dashed lines), $Q_{stat}/Q_*$ (short-dashed lines) and $Q/Q_{stat}$ (solid lines) as a function of matter density $\Omega_{m,0}$. Heavier lines are calculated for $z_f = 3$ while lighter ones have $z_f = \infty$.

are calculated for $z_f = 5$, while medium-weight (long-dashed) lines assume $z_f = 10$ and the lightest (short-dashed) lines have $z_f = 50$. Also shown is the boundary between big bang and bounce models (heavy solid line in top left corner), and the boundary between open and closed models (diagonal dashed line).

Fig. 2.6 shows that the bolometric intensity of the EBL is only modestly sensitive to the cosmological parameters $\Omega_{m,0}$ and $\Omega_{\Lambda,0}$. Moving from the lower right-hand corner of the phase space ($Q/Q_* = 0.37$) to the upper left-hand one ($Q/Q_* = 0.69$) changes the value of this quantity by less than a factor of two. Increasing the redshift of galaxy formation from $z_f = 5$ to 10 has little effect, and increasing it again to $z_f = 50$ even less. This means that, regardless of the redshift at which galaxies actually form, *essentially all of the light reaching us from outside the Milky Way comes from galaxies at $z < 5$.*

While Fig. 2.6 confirms that the night sky is dark in any reasonable cosmological model, Fig. 2.7 shows *why*. It is a contour plot of $Q/Q_{stat}$, the value of which varies so little across the phase space that we have restricted the range of $z_f$-values to keep the diagram from getting too cluttered. The

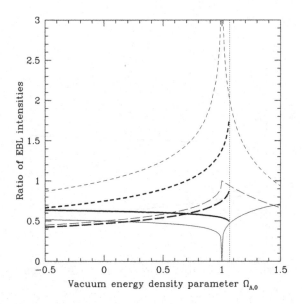

Fig. 2.5   Ratios $Q/Q_*$ (long-dashed lines), $Q_{stat}/Q_*$ (short-dashed lines) and $Q/Q_{stat}$ (solid lines) as a function of vacuum energy density $\Omega_{\Lambda,0}$. Heavier lines are calculated for $z_f = 3$ while lighter ones have $z_f = \infty$ for $\Omega_{\Lambda,0} \leqslant 1$ and $z_f = z_{max}$ for $\Omega_{\Lambda,0} > 1$. The dotted vertical line marks the maximum value of $\Omega_{\Lambda,0}$ for which one can integrate to $z_f = 3$.

heavy (solid) lines are calculated for $z_f = 4.5$, the medium-weight (long-dashed) lines for $z_f = 5$, and the lightest (short-dashed) lines for $z_f = 5.5$. The spread in contour values is extremely narrow, from $Q/Q_{stat} = 0.56$ in the upper left-hand corner to 0.64 in the lower right-hand one—a difference of less than 15%. Fig. 2.7 confirms our previous analytical results and leaves no doubt about the resolution of Olbers' paradox: *the brightness of the night sky is determined to order of magnitude by the lifetime of the galaxies, and reduced by a factor of only $0.6 \pm 0.1$ due to the expansion of the Universe.*

## 2.7   A bright sky at night?

In this section, we inquire into the evolution of bolometric EBL intensity $Q(t)$ with time, as specified by Eq. (2.10). To evaluate this integral, we require the form of the function $R(t)$, which can be expressed analytically if the Universe is 3D-flat, as suggested by observations of the cosmic mi-

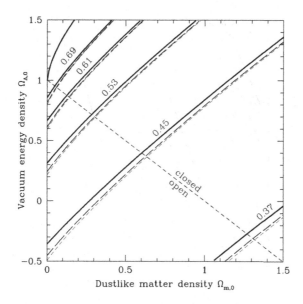

Fig. 2.6    The ratio $Q/Q_*$ in an expanding Universe, plotted as a function of matter density parameter $\Omega_{m,0}$ and vacuum density parameter $\Omega_{\Lambda,0}$ (with the radiation density parameter $\Omega_{r,0}$ set to zero). Solid lines correspond to $z_f = 5$, while long-dashed lines assume $z_f = 10$ and short-dashed ones have $z_f = 50$.

crowave background radiation (Chap. 4). For this case $k = 0$ and (2.20) simplifies to

$$\left(\frac{\dot{R}}{R}\right)^2 = \frac{8\pi G}{3}\left(\rho_r + \rho_m + \rho_\Lambda\right) , \qquad (2.47)$$

where we have taken $\rho = \rho_r + \rho_m$ in general and used (2.21) for $\rho_\Lambda$. If one of these three components is dominant at a given time, then we can make use of (2.30) to obtain

$$\left(\frac{\dot{R}}{R}\right)^2 = \frac{8\pi G}{3} \times \begin{cases} \rho_{r,0}(R/R_0)^{-4} & \text{(radiation)} \\ \rho_{m,0}(R/R_0)^{-3} & \text{(matter)} \\ \rho_\Lambda & \text{(vacuum)} \end{cases} . \qquad (2.48)$$

These differential equations are solved to give

$$\frac{R(t)}{R_0} = \begin{cases} (t/t_0)^{1/2} & \text{(radiation)} \\ (t/t_0)^{2/3} & \text{(matter)} \\ \exp[H_0(t - t_0)] & \text{(vacuum)} \end{cases} . \qquad (2.49)$$

We emphasize that these expressions assume spatial flatness and a single-component cosmic fluid, which must have the critical density.

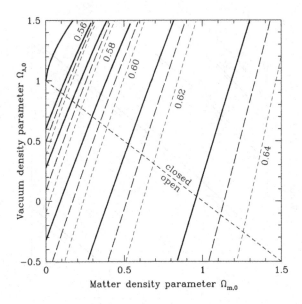

Fig. 2.7 The ratio $Q/Q_{\text{stat}}$ of EBL intensity in an expanding Universe to that in a static Universe with the same values of the matter density parameter $\Omega_{m,0}$ and vacuum density parameter $\Omega_{\Lambda,0}$ (with the radiation density parameter $\Omega_{r,0}$ set to zero). Solid lines correspond to $z_f = 4.5$, while long-dashed lines assume $z_f = 5$ and short-dashed ones have $z_f = 5.5$.

Putting (2.49) into (2.10), we can solve for the bolometric intensity under the assumption that the luminosity of the galaxies is constant over their lifetimes, $L(t) = L_0$:

$$\frac{Q(t)}{Q_*} = \begin{cases} (1/3)(t/t_0)^{3/2} & \text{(radiation)} \\ (2/5)(t/t_0)^{5/3} & \text{(matter)} \\ \exp\left[H_0 t_0(t/t_0 - 1)\right] & \text{(vacuum)} \end{cases}, \qquad (2.50)$$

where we have used (2.19) and assumed that $t_f \ll t_0$ and $t_f \ll t$.

The intensity of the light reaching us from intergalactic space climbs as $t^{3/2}$ in a radiation-filled Universe, $t^{5/3}$ in a matter-dominated one, and $\exp(H_0 t)$ in one which contains only vacuum energy. This happens because the horizon of the Universe expands to encompass more and more galaxies, and hence more photons. Clearly it does so at a rate which more than compensates for the dilution and redshifting of existing photons due to expansion. Suppose for argument's sake that this state of affairs could continue indefinitely. How long would it take for the night sky to become as bright as, say, the interior of a typical living-room containing a single 100 W bulb and a summed surface area of 100 m$^2$ ($Q_{\text{bulb}} = 10^3$ erg s$^{-1}$ cm$^{-2}$)? The required increase of $Q(t)$ over $Q_*$ ($= 2.5 \times 10^{-4}$ erg s$^{-1}$ cm$^{-2}$) is 4 million times. Eq. (2.50) then implies that

$$t \approx \begin{cases} 730\,000 \text{ Gyr} & \text{(radiation)} \\ 220\,000 \text{ Gyr} & \text{(matter)} \\ 230 \text{ Gyr} & \text{(vacuum)} \end{cases}, \qquad (2.51)$$

where we have taken $H_0 t_0 \approx 1$ and $t_0 \approx 14$ Gyr, as suggested by the observational data (Chap. 4). The last of the numbers in (2.51) is particularly intriguing. In a vacuum-dominated Universe in which the luminosity of the galaxies could be kept constant indefinitely, the night sky would fill up with light over timescales of the same order as the theoretical hydrogen-burning lifetimes of the longest-lived stars. Of course, the luminosity of galaxies *cannot* stay constant over these timescales, because much of their light comes from more massive stars which burn themselves out after tens of Gyr or less.

To check whether the situation just described is merely an artefact of the empty de Sitter Universe, we make use of an expression for $R(t)$ in 3D-flat models containing both matter and vacuum energy [1, 39, 46, 48, 49]:

$$\tilde{R}(t) = \left[ \sqrt{\frac{\Omega_{m,0}}{1 - \Omega_{m,0}}} \sinh\left(\tfrac{3}{2}\sqrt{1 - \Omega_{m,0}}\, H_0 t\right) \right]^{2/3} . \qquad (2.52)$$

Differentiating this with respect to time gives the Hubble expansion rate:

$$\tilde{H}(t) = \sqrt{1 - \Omega_{m,0}} \, \coth\left(\tfrac{3}{2}\sqrt{1 - \Omega_{m,0}}\, H_0 t\right) . \qquad (2.53)$$

This goes over to $\Omega_{\Lambda,0}^{1/2}$ as $t \to \infty$, a result which (as noted in Sec. 2.3) holds quite generally for models with $\Lambda > 0$. Alternatively, setting $\tilde{R} = (1+z)^{-1}$ in (2.52) gives the age of the Universe at redshift $z$:

$$t = \frac{2}{3H_0\sqrt{1 - \Omega_{m,0}}} \sinh^{-1}\sqrt{\frac{1 - \Omega_{m,0}}{\Omega_{m,0}(1+z)^3}} . \qquad (2.54)$$

Setting $z = 0$ in this equation gives the present age of the Universe ($t_0$). Thus a model with (say) $\Omega_{m,0} = 0.3$ and $\Omega_{\Lambda,0} = 0.7$ has an age of $t_0 = 0.96 H_0^{-1}$ or, using (2.16), $t_0 = 9.5 h_0^{-1}$ Gyr. Alternatively, in the limit $\Omega_{m,0} \to 1$, Eq. (2.54) gives back the standard result for the age of an EdS Universe, $t_0 = 2/(3H_0) = 6.5 h_0^{-1}$ Gyr.

Putting (2.52) into Eq. (2.10) with $L(t) = L_0 = $ constant, and integrating over time, we obtain the plots of bolometric EBL intensity shown in

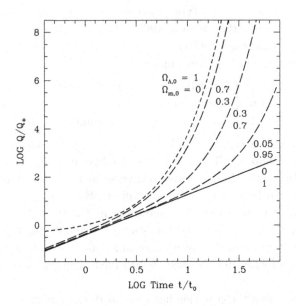

Fig. 2.8   Plots of $Q(t)/Q_*$ in 3D-flat models containing both dust-like matter ($\Omega_{m,0}$) and vacuum energy ($\Omega_{\Lambda,0}$). The solid line is the Einstein-de Sitter model, while the short-dashed line is pure de Sitter, and long-dashed lines represent intermediate models. The curves do not meet at $t = t_0$ because $Q(t_0)$ differs from model to model, with $Q(t_0) = Q_*$ only for the pure de Sitter case.

Fig. 2.8. This diagram shows that the specter of a rapidly-brightening Universe does not occur only in the pure de Sitter model (short-dashed line). A model with an admixture of dust-like matter and vacuum energy such that $\Omega_{m,0} = 0.3$ and $\Omega_{\Lambda,0} = 0.7$, for instance, takes only modestly longer (280 Gyr) to attain a "living-room" intensity of light.

In theory then, it might be thought that our remote descendants could live under skies perpetually ablaze with the light of distant galaxies, in which the rising and setting of their home suns would barely be noticed. Of course, this will not happen in practice because galaxy luminosities change with time as their brightest stars burn out. Thus the lifetime of the galaxies is again critical: it determines the brightness of the night sky not only in the sense of the past, but the future as well.

Chapter 3

# The Spectrum of Cosmic Background Light

## 3.1 Spectral intensity

The spectra of real galaxies depend strongly on wavelength and also evolve with time. How might these facts alter the conclusion obtained in Chap. 2; namely, that the brightness of the night sky is overwhelmingly determined by the age of the Universe, with expansion playing only a minor role?

The significance of this question is best appreciated in the microwave portion of the electromagnetic spectrum (at wavelengths from about 1 mm to 10 cm) where we know from decades of radio astronomy that the "night sky" is brighter than its optical counterpart (Fig. 1.15). The majority of this microwave background radiation is thought to come, not from the redshifted light of distant galaxies, but from the fading glow of the big bang itself—the "ashes and smoke" of creation in Lemaître's words. Since its nature and suspected origin are different from those of the EBL, this part of the spectrum has its own name, the *cosmic microwave background* (CMB). Here expansion is of paramount importance, since the source radiation in this case was emitted at more or less a single instant in cosmological history (so that the "lifetime of the sources" is negligible). Another way to see this is to take expansion out of the picture, as we did in Sec. 2.4: the CMB intensity we would observe in this "equivalent static model" would be that of the primordial fireball and would roast us alive.

While Olbers' paradox involves the EBL, not the CMB, this example is still instructive because it prompts us to consider whether similar (though less pronounced) effects could have been operative in the EBL as well. If, for instance, galaxies emitted most of their light in a relatively brief burst of star formation at very early times, this would be a galactic approximation to the picture just described, and could conceivably boost the importance of expansion relative to lifetime, at least in some wavebands. To check on this, we need a way to calculate EBL intensity as a function of wavelength. This is motivated by other considerations as well. Olbers' paradox has historically been concerned primarily with the *optical* waveband (from approximately 4000Å to 8000Å), and this is still what most people mean when they refer to the "brightness of the night sky." And from a practical standpoint, we would like to compare our theoretical predictions with observational data, and these are necessarily taken using detectors which are optimized for finite portions of the electromagnetic spectrum.

We therefore generalize the bolometric formalism of Chap. 2. Instead of total luminosity $L$, consider the energy emitted by a source per unit time between wavelengths $\lambda$ and $\lambda + d\lambda$. Let us write this in the form $dL_\lambda \equiv F(\lambda, t) \, d\lambda$ where $F(\lambda, t)$ is the *spectral energy distribution* (SED), with dimensions of energy per unit time per unit wavelength. Luminosity is recovered by integrating the SED over all wavelengths:

$$L(t) = \int_0^\infty dL_\lambda = \int_0^\infty F(\lambda, t) \, d\lambda \, . \tag{3.1}$$

We then return to (2.9), the bolometric intensity of the spherical shell of galaxies depicted in Fig. 2.1. Replacing $L(t)$ with $dL_\lambda$ in this equation gives the intensity of light emitted between $\lambda$ and $\lambda + d\lambda$:

$$dQ_{\lambda,\text{em}} = cn(t)\tilde{R}(t)[F(\lambda, t) \, d\lambda]dt \, . \tag{3.2}$$

This light reaches us at the redshifted wavelength $\lambda_0 = \lambda/\tilde{R}(t)$. Redshift also stretches the wavelength interval by the same factor, $d\lambda_0 = d\lambda/\tilde{R}(t)$. So the intensity of light *observed* by us between $\lambda_0$ and $\lambda_0 + d\lambda_0$ is

$$dQ_{\lambda,\text{obs}} = cn(t)\tilde{R}^2(t)F[\tilde{R}(t)\lambda_0, t] \, d\lambda_0 dt \, . \tag{3.3}$$

The intensity of the shell *per unit wavelength*, as observed at wavelength $\lambda_0$, is then given simply by

$$4\pi \, dI_\lambda(\lambda_0) \equiv \frac{dQ_{\lambda,\text{obs}}}{d\lambda_0} = cn(t)\tilde{R}^2(t)F[\tilde{R}(t)\lambda_0, t] \, d\lambda_0 dt \, , \tag{3.4}$$

where the factor $4\pi$ converts from an all-sky intensity to one measured per steradian. (This is merely a convention, but has become standard.) Integrating over all the spherical shells corresponding to times $t_0$ and $t_0 - t_f$ (as before) we obtain the spectral analog of our earlier bolometric result, Eq. (2.10):

$$I_\lambda(\lambda_0) = \frac{c}{4\pi} \int_{t_f}^{t_0} n(t)F[\tilde{R}(t)\lambda_0, t]\tilde{R}^2(t) \, dt \, . \tag{3.5}$$

This is the integrated light from many galaxies, which has been emitted at various wavelengths and redshifted by various amounts, but which is all in the waveband centered on $\lambda_0$ when it arrives at us. We refer to this as the *spectral intensity of the EBL* at $\lambda_0$. Eq. (3.5), or ones like it, have been considered from the theoretical side principally by G.C. McVittie and S.P. Wyatt [50], G.J. Whitrow and B.D. Yallop [51, 52] and P.S. Wesson *et al.* [36, 37].

Eq. (3.5) can be converted from an integral over $t$ to one over $z$ by means of Eq. (2.12) as before. This gives

$$I_\lambda(\lambda_0) = \frac{c}{4\pi H_0} \int_0^{z_f} \frac{n(z)F[\lambda_0/(1+z), z]\,dz}{(1+z)^3 \tilde{H}(z)} \, . \tag{3.6}$$

Eq. (3.6) is the spectral analog of (2.13). It may be checked using (3.1) that bolometric intensity is just the integral of spectral intensity over all observed wavelengths, $Q = \int_0^\infty I(\lambda_0)d\lambda_0$. Eqs. (3.5) and (3.6) provide us with the means to constrain any kind of radiation source by means of its contributions to the background light, once its number density $n(z)$ and energy spectrum $F(\lambda, z)$ are known. In subsequent chapters we will apply them to various species of dark (and not so dark) energy and matter.

In this chapter, we return to the question of Olbers' paradox. The static analog of Eq. (3.5) (i.e. the equivalent spectral EBL intensity in a universe without expansion, but with the properties of the galaxies unchanged) is obtained exactly as in the bolometric case by setting $\tilde{R}(t) = 1$ (Sec. 2.4):

$$I_{\lambda,\text{stat}}(\lambda_0) = \frac{c}{4\pi} \int_{t_f}^{t_0} n(t)F(\lambda_0, t)\,dt \, . \tag{3.7}$$

Just as before, we may convert this to an integral over $z$ if we choose. The latter parameter no longer represents physical redshift (since this has been eliminated by hypothesis), but is now merely an algebraic way of expressing the age of the galaxies. This is convenient because it puts (3.7) into a form which may be directly compared with its counterpart (3.6) in the expanding Universe:

$$I_{\lambda,\text{stat}}(\lambda_0) = \frac{c}{4\pi H_0} \int_0^{z_f} \frac{n(z)F(\lambda_0, z)\,dz}{(1+z)\tilde{H}(z)} \, . \tag{3.8}$$

If the same values are adopted for $H_0$ and $z_f$, and the same functional forms are used for $n(z), F(\lambda, z)$ and $\tilde{H}(z)$, then Eqs. (3.6) and (3.8) allow us to compare model universes which are alike in every way, except that one is expanding while the other stands still.

Some simplification of these expressions is obtained as before in situations where the comoving source number density can be taken as constant,

$n(z) = n_0$. However, it is not possible to go farther and pull all the dimensional content out of these integrals, as was done in the bolometric case, until a specific form is postulated for the SED $F(\lambda, z)$.

## 3.2  Luminosity density

The simplest possible source spectrum is one in which all the energy is emitted at a single peak wavelength $\lambda_p$ at each redshift $z$. We can write this in terms of the Dirac $\delta$-function as

$$F(\lambda, z) = F_p(z)\,\delta\left(\frac{\lambda}{\lambda_p} - 1\right) . \tag{3.9}$$

SEDs of this form are well-suited to sources of electromagnetic radiation such as elementary particle decays, which are characterized by specific decay energies and may occur in the dark-matter halos surrounding galaxies. The $\delta$-function SED is not a realistic description of galaxy spectra, but we apply it in this context here as a way of laying the foundation for subsequent sections.

The function $F_p(z)$ is obtained in terms of the total source luminosity $L(z)$ by normalizing over all wavelengths via

$$L(z) \equiv \int_0^\infty F(\lambda, z)\,d\lambda = F_p(z)\lambda_p , \tag{3.10}$$

so that $F_p(z) = L(z)/\lambda_p$. In the case of galaxies, a logical choice for the characteristic wavelength $\lambda_p$ would be the peak wavelength of a blackbody of "typical" stellar temperature. Taking the Sun as typical ($T = T_\odot = 5770$K), this would be $\lambda_p = (2.90$ mm K$)/T = 5020$Å from Wiens' law. Distant galaxies are seen chiefly during periods of intense starburst activity when many stars are much hotter than the Sun, suggesting a shift toward shorter wavelengths. On the other hand, most of the short-wavelength light produced in large starbursting galaxies (as much as 99% in the most massive cases) is absorbed within these galaxies by dust and re-radiated in the infrared and microwave regions ($\lambda \gtrsim 10,000$Å). It is also important to keep in mind that while distant starburst galaxies may be hotter and more luminous than local spirals and ellipticals, the latter contribute most to EBL intensity by virtue of their numbers at low redshift. The best that one can do with a single characteristic wavelength is to locate it somewhere within the B-band ($3600 - 5500$Å). For the purposes of this exercise we associate $\lambda_p$ with the nominal center of this band, $\lambda_p = 4400$Å, corresponding to a blackbody temperature of 6590 K.

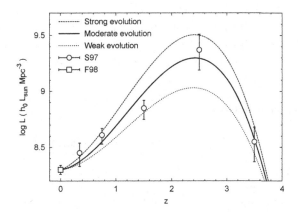

Fig. 3.1   The comoving luminosity density of the Universe $\mathcal{L}(z)$ (in $h_0 L_\odot$ Mpc$^{-3}$), as observed at $z = 0$ (square) and extrapolated to higher redshifts based on analysis of the Hubble Deep Field (circles). The solid curve is a least-squares fit to the data; dashed lines represent upper and lower limits.

Substituting the SED (3.9) into Eq. (3.6) leads to

$$I_\lambda(\lambda_0) = \frac{c}{4\pi H_0 \lambda_p} \int_0^{z_f} \frac{\mathcal{L}(z)}{(1+z)^3 \tilde{H}(z)} \, \delta\left[ \frac{\lambda_0}{\lambda_p(1+z)} - 1 \right] dz \,, \qquad (3.11)$$

where we have introduced a new shorthand for the *comoving luminosity density* of galaxies:

$$\mathcal{L}(z) \equiv n(z)L(z) \,. \qquad (3.12)$$

At redshift $z = 0$ this takes the value $\mathcal{L}_0$, as given by (2.18). Numerous studies have shown that the product of $n(z)$ and $L(z)$ is approximately conserved with redshift, even when the two quantities themselves appear to be evolving markedly. So it would be reasonable to take $\mathcal{L}(z) = \mathcal{L}_0 = $ const. However, recent analyses at deeper redshifts indicate that $\mathcal{L}(z)$ does rise slowly but steadily with $z$, peaking in the range $2 \lesssim z \lesssim 3$, and falling away sharply thereafter [53]. This is consistent with a picture in which the first generation of galaxy formation occurred near $z \sim 3$, being followed at lower redshifts by galaxies whose evolution proceeded more passively.

Fig. 3.1 shows the value of $\mathcal{L}_0$ from (2.18) at $z = 0$ [40] with an extrapolation of $\mathcal{L}(z)$ to higher redshifts from an analysis of photometric galaxy redshifts in the Hubble Deep Field (HDF) [54]. We define a *relative comoving luminosity density* $\tilde{\mathcal{L}}(z)$ by

$$\tilde{\mathcal{L}}(z) \equiv \mathcal{L}(z)/\mathcal{L}_0 \,, \qquad (3.13)$$

and fit this to the data with a cubic $[\log \tilde{\mathcal{L}}(z) = \alpha z + \beta z^2 + \gamma z^3]$. The best least-squares fit is plotted as a solid line in Fig. 3.1 along with upper and lower limits (dashed lines). We refer to these cases in what follows as the "moderate," "strong" and "weak" galaxy-evolution scenarios respectively.

## 3.3 The delta function

Inserting (3.13) into (3.11) puts the latter into the form

$$I_\lambda(\lambda_0) = I_\delta \int_0^{z_f} \frac{\tilde{\mathcal{L}}(z)}{(1+z)^3 \tilde{H}(z)} \delta \left[ \frac{\lambda_0}{\lambda_p(1+z)} - 1 \right] dz . \tag{3.14}$$

The dimensional content of this integral has been concentrated into a prefactor $I_\delta$, defined by

$$I_\delta = \frac{c\mathcal{L}_0}{4\pi H_0 \lambda_p} = 4.4 \times 10^{-9} \text{erg s}^{-1} \text{ cm}^{-2} \text{ Å}^{-1} \text{ ster}^{-1} \left( \frac{\lambda_p}{4400\text{Å}} \right)^{-1}. \tag{3.15}$$

This constant shares two important properties of its bolometric counterpart $Q_*$ (Sec. 2.2). First, it is independent of the uncertainty $h_0$ in Hubble's constant. Second, it is *low* by everyday standards. It is, for example, far below the intensity of the zodiacal light, which is caused by the scattering of sunlight by dust in the plane of the solar system. This is important, since the value of $I_\delta$ sets the scale of the integral (3.14). In fact, observational upper bounds on $I_\lambda(\lambda_0)$ at $\lambda_0 \approx 4400$Å are typically of the same order as $I_\delta$. The Pioneer 10 photopolarimeter, for instance, was used to set one such limit of $I_\lambda(4400\text{Å}) < 4.5 \times 10^{-9} \text{erg s}^{-1} \text{ cm}^{-2} \text{ Å}^{-1} \text{ ster}^{-1}$ [55].

Dividing $I_\delta$ of (3.15) by the photon energy $E_0 = hc/\lambda_0$ (where $hc = 1.986 \times 10^{-8}$ erg Å) puts the EBL intensity integral (3.14) into new units, sometimes referred to as *continuum units* (CUs):

$$I_\delta = I_\delta(\lambda_0) = \frac{\mathcal{L}_0}{4\pi h H_0} \left( \frac{\lambda_0}{\lambda_p} \right) = 970 \text{ CUs} \left( \frac{\lambda_0}{\lambda_p} \right) , \tag{3.16}$$

where 1 CU $\equiv$ 1 photon s$^{-1}$ cm$^{-2}$ Å$^{-1}$ ster$^{-1}$. While both kinds of units (CUs and erg s$^{-1}$ cm$^{-2}$ Å$^{-1}$ ster$^{-1}$) are in common use for reporting spectral intensity at near-optical wavelengths, CUs appear most frequently. They are also preferable from a theoretical point of view, because they most faithfully reflect the *energy content* of a spectrum [56]. A third type of intensity unit, the $S_{10}$ (loosely, the equivalent of one tenth-magnitude star per square degree) is also occasionally encountered but will be avoided in this book as it is wavelength-dependent and involves other subtleties which differ between workers.

Table 3.1 Cosmological test models

|  | EdS/SCDM | OCDM | ΛCDM | ΛBDM |
| --- | --- | --- | --- | --- |
| $\Omega_{m,0}$ | 1 | 0.3 | 0.3 | 0.03 |
| $\Omega_{\Lambda,0}$ | 0 | 0 | 0.7 | 1 |

If we let the redshift of formation $z_f \to \infty$ then Eq. (3.14) reduces to

$$I_\lambda(\lambda_0) = \begin{cases} I_\delta \left(\dfrac{\lambda_0}{\lambda_p}\right)^{-2} \dfrac{\tilde{\mathcal{L}}(\lambda_0/\lambda_p - 1)}{\tilde{H}(\lambda_0/\lambda_p - 1)} & (\text{if } \lambda_0 \geqslant \lambda_p) \\ 0 & (\text{if } \lambda_0 < \lambda_p) \end{cases} \qquad (3.17)$$

The comoving luminosity density $\tilde{\mathcal{L}}(\lambda_0/\lambda_p - 1)$ which appears here is fixed by the fit (3.13) to the HDF data in Fig. 3.1. The Hubble parameter is given by (2.31) as $\tilde{H}(\lambda_0/\lambda_p - 1) = [\Omega_{m,0}(\lambda_0/\lambda_p)^3 + \Omega_{\Lambda,0} - (\Omega_{m,0} + \Omega_{\Lambda,0} - 1)(\lambda_0/\lambda_p)^2]^{1/2}$ for a universe containing dust-like matter and vacuum energy with density parameters $\Omega_{m,0}$ and $\Omega_{\Lambda,0}$ respectively.

Turning off the luminosity density evolution (so that $\tilde{\mathcal{L}} = 1 = $ const.), one obtains three simple special cases:

$$I_\lambda(\lambda_0) = I_\delta \times \begin{cases} (\lambda_0/\lambda_p)^{-7/2} & (\text{Einstein-de Sitter}) \\ (\lambda_0/\lambda_p)^{-2} & (\text{de Sitter}) \\ (\lambda_0/\lambda_p)^{-3} & (\text{Milne}) \end{cases} \qquad (3.18)$$

These are taken at $\lambda_0 \geqslant \lambda_p$, where $(\Omega_{m,0}, \Omega_{\Lambda,0}) = (1,0), (0,1)$ and $(0,0)$ respectively for the three models cited (Table 2.1). The first of these is the "$\frac{7}{2}$-law" which often appears in the particle-physics literature as an approximation to the spectrum of EBL contributions from decaying particles. But the second (de Sitter) probably provides a better approximation, given current thinking regarding the values of $\Omega_{m,0}$ and $\Omega_{\Lambda,0}$.

To evaluate the spectral EBL intensity (3.14) and other quantities in a general situation, it will be helpful to define a suite of *cosmological test models* which span the widest possible range in the parameter space defined by $\Omega_{m,0}$ and $\Omega_{\Lambda,0}$ (Table 3.1). The Einstein-de Sitter (EdS) model has long been favored on grounds of simplicity, and is still sometimes referred to as the "standard cold dark matter" or SCDM model. It has come under increasing pressure, however, as evidence mounts for levels of $\Omega_{m,0} \lesssim 0.5$, and most recently from observations of Type Ia supernovae (SNIa) which indicate that $\Omega_{\Lambda,0} > \Omega_{m,0}$. The *Open Cold Dark Matter* (OCDM) model is more consistent with data on $\Omega_{m,0}$ and holds appeal for those who have been reluctant to accept the possibility of a finite vacuum energy. It faces the considerable challenge, however, of explaining data on the spectrum of CMB fluctuations, which imply that $\Omega_{m,0} + \Omega_{\Lambda,0} \approx 1$. The $\Lambda + Cold\ Dark$ *Matter* ($\Lambda$CDM) model has rapidly become the new standard in cosmology because it agrees best with both the SNIa and CMB observations. However, this model suffers from a "coincidence problem," in that $\Omega_m(t)$ and $\Omega_\Lambda(t)$ evolve so differently with time that the probability of finding ourselves at a moment in cosmic history when they are even of the same order of magnitude appears unrealistically small. This is addressed to some extent in the last model, where we push $\Omega_{m,0}$ and $\Omega_{\Lambda,0}$ to their lowest and highest limits, respectively. In the case of $\Omega_{m,0}$ these limits are set by big-bang nucleosynthesis, which requires a density of at least $\Omega_{m,0} \approx 0.03$ in baryons (hence the $\Lambda + Baryonic\ Dark\ Matter$ or $\Lambda$BDM model). Upper limits on $\Omega_{\Lambda,0}$ come from the frequency of gravitational lenses and the requirement that the Universe began in a big-bang singularity. Within the context of isotropic and homogeneous cosmology, these four models cover the full range of what would be considered plausible by most workers.

Fig. 3.2 shows the solution of the full integral (3.14) for all four test models, superimposed on a plot of available experimental data at near-optical wavelengths (i.e. a close-up of Fig. 1.15). The short-wavelength cutoff in these plots is an artefact of the $\delta$-function SED, but the behavior of $I_\lambda(\lambda_0)$ at wavelengths above $\lambda_p = 4400$ Å is more revealing. In the EdS case (a), the rapid fall-off in intensity with $\lambda_0$ indicates that *nearby* (low-redshift) galaxies dominate. There is a secondary hump at $\lambda_0 \approx 10,000$ Å, which is an "echo" of the peak in galaxy formation, redshifted into the near infrared. This hump becomes progressively larger relative to the optical peak at 4400 Å as the ratio of $\Omega_{\Lambda,0}$ to $\Omega_{m,0}$ grows. Eventually one has the situation in the de Sitter-like model (d), where the galaxy-formation peak entirely dominates the observed EBL signal, despite the fact that it comes from distant galaxies at $z \approx 3$. This is because a large $\Omega_{\Lambda,0}$-term (especially one which is large relative to $\Omega_{m,0}$) inflates comoving volume at high redshifts. Since the comoving *number density* of galaxies is fixed by the fit to observational data on $\tilde{\mathcal{L}}(z)$ (Fig. 3.1), the number of galaxies at these redshifts must go up, pushing up the infrared part of the spectrum. We will see that this trend persists in more sophisticated models, providing a clear link between observations of the EBL and the cosmological parameters $\Omega_{m,0}$ and $\Omega_{\Lambda,0}$.

Fig. 3.2 is plotted over a broad range of wavelengths from the near ultraviolet (NUV; 2000–4000Å) to the near infrared (NIR; 8000–40,000Å). The upper limits in this plot (solid symbols and heavy lines) come from

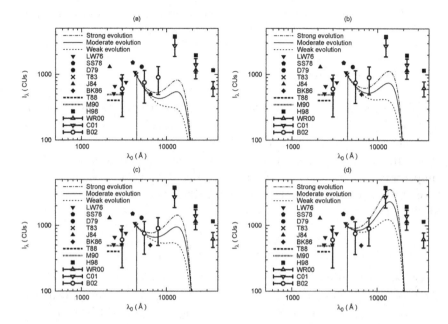

Fig. 3.2 The spectral EBL intensity of galaxies whose radiation is modelled by δ-functions at a rest frame wavelength of 4400Å, calculated for four different cosmological models: (a) EdS, (b) OCDM, (c) ΛCDM and (d) ΛBDM (Table 3.1). Also shown are observational upper limits (solid symbols and heavy lines) and reported detections (empty symbols) over the waveband 2000–40,000Å.

analyses of OAO-2 satellite data (LW76 [57]), ground-based telescopes (SS78 [58], D79 [59], BK86 [60]), Pioneer 10 (T83 [55]), sounding rockets (J84 [61], T88 [62]), the shuttle-borne Hopkins UVX (M90 [63]) and—in the near infrared—the DIRBE instrument aboard the COBE satellite (H98 [64]). Recent years have also seen the first widely-accepted *detections* of the EBL (Fig. 3.2, open symbols). In the NIR these have come from continued analysis of DIRBE data in the K-band (22,000Å) and L-band (35,000Å; WR00 [65]), as well as the J-band (12,500Å; C01 [66]). Reported detections in the optical using a combination of Hubble Space Telescope (HST) and Las Campanas telescope observations (B02 [67]) are preliminary [68] but potentially very important.

Fig. 3.2 shows that EBL intensities based on the simple δ-function spectrum are in rough agreement with these data. Predicted intensities come in at or just below the optical limits in the low-$\Omega_{\Lambda,0}$ cases (a) and (b), and remain consistent with the infrared limits even in the high-$\Omega_{\Lambda,0}$ cases

(c) and (d). Vacuum-dominated models with even higher ratios of $\Omega_{\Lambda,0}$ to $\Omega_{m,0}$ would, however, run afoul of DIRBE limits in the J-band.

## 3.4   The normal distribution

The Gaussian distribution provides a useful generalization of the $\delta$-function for modelling sources whose spectra, while essentially monochromatic, are broadened by some physical process (such as a Doppler shift due to rotation). In the context of galaxies, this extra degree of freedom provides a simple way to model the width of the bright part of the spectrum. If we take this to cover the B-band (3600-5500Å) then $\sigma_\lambda \sim 1000$Å. The Gaussian SED reads

$$F(\lambda, z) = \frac{L(z)}{\sqrt{2\pi}\,\sigma_\lambda} \exp\left[-\frac{1}{2}\left(\frac{\lambda - \lambda_p}{\sigma_\lambda}\right)^2\right], \qquad (3.19)$$

where $\lambda_p$ is the wavelength at which the galaxy emits most of its light. We take $\lambda_p = 4400$Å as before, and note that integration over $\lambda_0$ confirms that $L(z) = \int_0^\infty F(\lambda, z)d\lambda$ as required. Using the empirical fit to $\tilde{\mathcal{L}}(z) \equiv n(z)L(z)/\mathcal{L}_0$ from the HDF data in Fig. 3.1 and substituting the SED (3.19) into Eq. (3.6), we obtain

$$I_\lambda(\lambda_0) = I_g \int_0^{z_f} \frac{\tilde{\mathcal{L}}(z)}{(1+z)^3 \tilde{H}(z)} \exp\left\{-\frac{1}{2}\left[\frac{\lambda_0/(1+z) - \lambda_p}{\sigma_\lambda}\right]^2\right\} dz. \quad (3.20)$$

The dimensional content of this integral has been pulled into a prefactor $I_g = I_g(\lambda_0)$, defined by

$$I_g = \frac{\mathcal{L}_0}{\sqrt{32\pi^3}\,hH_0}\left(\frac{\lambda_0}{\sigma_\lambda}\right) = 390 \text{ CUs}\left(\frac{\lambda_0}{\sigma_\lambda}\right). \qquad (3.21)$$

Here we have divided (3.20) by the photon energy $E_0 = hc/\lambda_0$ to put the result into CUs, as before.

Results are shown in Fig. 3.3, where we have taken $\lambda_p = 4400$Å, $\sigma_\lambda = 1000$Å and $z_f = 6$. Aside from the fact that the short-wavelength cutoff has disappeared, the situation is qualitatively similar to that obtained using a $\delta$-function approximation. As before, the expected EBL signal is brightest at optical wavelengths in an EdS Universe (a), but the infrared hump due to the redshifted peak of galaxy formation begins to dominate for higher-$\Omega_{\Lambda,0}$ models (b) and (c), becoming overwhelming in the de Sitter-like model (d). Overall, the best agreement between calculated and observed EBL levels occurs in the $\Lambda$CDM model (c). The matter-dominated EdS (a) and OCDM

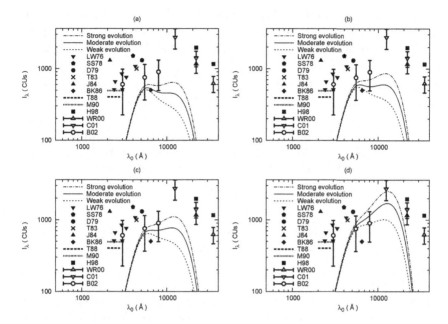

Fig. 3.3 The spectral EBL intensity of galaxies whose spectra has been represented by Gaussian distributions with rest frame peak wavelength 4400Å and standard deviation 1000Å, calculated for the (a) EdS, (b) OCDM, (c) ΛCDM and (d) ΛBDM cosmologies and compared with observational upper limits (solid symbols and heavy lines) and reported detections (empty symbols).

(b) models contain too little light (implying an additional source of optical or near-optical background radiation besides that from galaxies), while the ΛBDM model (d) comes uncomfortably close to containing *too much* light. This is an interesting situation, and one which motivates us to reconsider the problem with more realistic models for the galaxy SED.

## 3.5 The thermal spectrum

The simplest nontrivial approach to a galaxy spectrum is to model it as a blackbody, and this was done by previous workers such as G.C. McVittie and S.P. Wyatt [50], G.J. Whitrow and B.D. Yallop [51, 52] and P.S. Wesson [37]. Let us suppose that the galaxy SED is a product of the Planck function and some wavelength-independent parameter $C(z)$:

$$F(\lambda, z) = \frac{2\pi hc^2}{\sigma_{\rm SB}} \frac{C(z)/\lambda^5}{\exp\left[hc/kT(z)\lambda\right] - 1}. \qquad (3.22)$$

Here $\sigma_{SB} \equiv 2\pi^5 k^4/15c^2 h^3 = 5.67 \times 10^{-5}$ erg s$^{-1}$ cm$^{-2}$ K$^{-1}$ is the Stefan-Boltzmann constant. The function $F$ is normally regarded as an increasing function of redshift (at least out to the redshift of galaxy formation). This is accommodated by allowing $C(z)$ or $T(z)$ to increase with $z$ in (3.22). The former choice implies that galaxy luminosities decrease with time while their spectrum remains unchanged, as might happen if stars were simply to disappear. The second implies that galaxy luminosities decrease with time as stars age and their spectrum becomes *redder*. The latter scenario is more realistic, and will be adopted here. The luminosity $L(z)$ is found by integrating $F(\lambda, z)$ over all wavelengths:

$$L(z) = \frac{2\pi hc^2}{\sigma_{SB}} C(z) \int_0^\infty \frac{\lambda^{-5}d\lambda}{\exp\left[hc/kT(z)\lambda\right] - 1} = C(z)\left[T(z)\right]^4 , \quad (3.23)$$

so that the unknown function $C(z)$ must satisfy $C(z) = L(z)/[T(z)]^4$. If we require that Stefan's law ($L \propto T^4$) hold at each $z$, then

$$C(z) = \text{const.} = L_0/T_0^4 , \quad (3.24)$$

where $T_0$ is the present "galaxy temperature" (i.e. the blackbody temperature corresponding to a peak wavelength in the B-band). Thus the evolution of galaxy luminosity in this model is just what is required by Stefan's law for blackbodies whose *temperatures* evolve as $T(z)$. This is reasonable, since galaxies are made up of stellar populations which cool and redden with time as hot massive stars die out.

Let us supplement this with the assumption of constant comoving number density, $n(z) = n_0 = $ const. This is sometimes referred to as the pure luminosity evolution or PLE scenario, and while there is some controversy on this point, PLE has been found by many workers to be roughly consistent with observed numbers of galaxies at faint magnitudes, especially if there is a significant vacuum energy density $\Omega_{\Lambda,0} > 0$. Proceeding on this assumption, the comoving galaxy luminosity density can be written

$$\tilde{\mathcal{L}}(z) \equiv \frac{n(z)L(z)}{\mathcal{L}_0} = \frac{L(z)}{L_0} = \left[\frac{T(z)}{T_0}\right]^4 . \quad (3.25)$$

This expression can then be inverted for blackbody temperature $T(z)$ as a function of redshift, since the form of $\tilde{\mathcal{L}}(z)$ is fixed by Fig. 3.1:

$$T(z) = T_0[\tilde{\mathcal{L}}(z)]^{1/4} . \quad (3.26)$$

We can check this by choosing $T_0 = 6600$K (i.e. a present peak wavelength of 4400Å) and reading off values of $\tilde{\mathcal{L}}(z) = \mathcal{L}(z)/\mathcal{L}_0$ at the peaks of the curves marked "weak," "moderate" and "strong" evolution in Fig. 3.1.

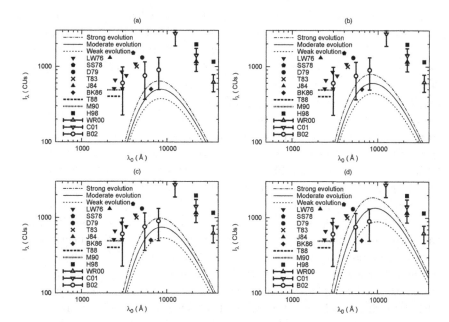

Fig. 3.4 The spectral EBL intensity of galaxies, modelled as blackbodies whose characteristic temperatures are such that their luminosities $L \propto T^4$ combine to produce the observed comoving luminosity density $\mathcal{L}(z)$ of the Universe. Results are shown for the (a) EdS, (b) OCDM, (c) $\Lambda$CDM and (d) $\Lambda$BDM cosmologies. Also shown are observational upper limits (solid symbols and heavy lines) and reported detections (open symbols).

Putting these numbers into (3.26) yields blackbody temperatures (and corresponding peak wavelengths) of 10,000K (2900Å), 11,900K (2440Å) and 13,100K (2210Å) respectively at the galaxy-formation peak. These numbers are consistent with the idea that galaxies would have been dominated by hot UV-emitting stars at this early time.

Inserting the expressions (3.24) for $C(z)$ and (3.26) for $T(z)$ into the SED (3.22), and substituting the latter into Eq. (3.6), we obtain

$$I_\lambda(\lambda_0) = I_b \int_0^{z_f} \frac{(1+z)^2 dz}{\{\exp\left[hc\,(1+z)/kT(z)\lambda_0\right] - 1\}\,\tilde{H}(z)} \,. \tag{3.27}$$

The dimensional prefactor $I_b = I_b(T_0, \lambda_0)$ reads in this case

$$I_b = \frac{c^2 \mathcal{L}_0}{2 H_0 \, \sigma_{\text{SB}} T_0^4 \lambda_0^4} = 90,100 \text{ CUs} \left(\frac{T_0}{6600\text{K}}\right)^{-4} \left(\frac{\lambda_0}{4400\text{Å}}\right)^{-4} \,. \tag{3.28}$$

This integral is evaluated and plotted in Fig. 3.4, assuming $z_f = 6$ as suggested by recent evidence for an epoch of "first light" at this redshift [69]. (Results are insensitive to this choice, provided that $z_f \gtrsim 3$.)

Fig. 3.4 shows some qualitative differences from our earlier results obtained using $\delta$-function and Gaussian SEDs. Most noticeably, the prominent "double-hump" structure is no longer apparent. The key evolutionary parameter is now blackbody temperature $T(z)$ and this goes as $[\mathcal{L}(z)]^{1/4}$ so that individual features in the comoving luminosity density profile are suppressed. As before, however, the long-wavelength part of the spectrum climbs steadily up the right-hand side of the figure as one moves from the $\Omega_{\Lambda,0} = 0$ models (a) and (b) to the $\Omega_{\Lambda,0}$-dominated models (c) and (d), whose light comes increasingly from more distant, redshifted galaxies.

Absolute EBL intensities in each of these four models are consistent with what we have seen already. This is not surprising, because changing the shape of the SED merely shifts light from one part of the spectrum to another. It cannot alter the *total amount* of light in the EBL, which is set by the comoving luminosity density $\tilde{\mathcal{L}}(z)$ of sources (once the background cosmology has been chosen). As before, the best match between calculated EBL intensities and the observational detections is found for the $\Omega_{\Lambda,0}$-dominated models (c) and (d). The fact that the EBL is now spread across a broader spectrum has slightly reduced its peak intensity, so that the best overall fit to the data now comes from the $\Lambda$BDM model (d). The zero-$\Omega_{\Lambda,0}$ models (a) and (b) again appear to require some additional source of background radiation (beyond that produced by galaxies) if they are to contain enough light to make up the levels of EBL intensity that have been reported.

## 3.6   The spectra of galaxies

The previous sections have shown that simple models of galaxy spectra, combined with data on the evolution of comoving luminosity density in the Universe, can produce levels of spectral EBL intensity in rough agreement with observational limits and reported detections, and even discriminate to a degree between different cosmological models. However, the results obtained up to this point are somewhat unsatisfactory in that they are sensitive to theoretical input parameters, such as $\lambda_p$ and $T_0$, which are hard to connect with the properties of the actual galaxy population.

A more comprehensive approach would use observational data in conjunction with theoretical models of galaxy evolution to build up an ensemble of evolving galaxy SEDs $F(\lambda, z)$ and comoving number densities $n(z)$ which would depend not only on redshift but on *galaxy type* as well. Increasingly sophisticated work has been carried out along these lines over the years by R.B. Partridge and P.J.E. Peebles [70], N.M. Tinsley [71], A.G. Bruzual [72], A.D. Code and G.A. Welch [73], Y. Yoshii and F. Takahara [74] and others. The last-named authors, for instance, divided galaxies into five morphological types (E/SO, Sab, Sbc, Scd and Sdm), with a different evolving SED for each type, and found that their collective EBL intensity at NIR wavelengths was about an order of magnitude below the levels suggested by observation.

Models of this kind, however, are complicated while at the same time containing uncertainties. This makes their use somewhat incompatible with our purpose here, which is to obtain a first-order estimate of EBL intensity so that the importance of *expansion* can be properly ascertained. Also, observations have begun to show that the above morphological classifications are of limited value at redshifts $z \gtrsim 1$, where spirals and ellipticals are still in the process of forming [75]. As we have already seen, this is precisely where much of the EBL may originate, especially if luminosity density evolution is strong, or if there is a significant $\Omega_{\Lambda,0}$-term.

What is needed, then, is a simple model which does not distinguish too finely between the spectra of galaxy types as they have traditionally been classified, but which can capture the essence of broad trends in luminosity density evolution over the full range of redshifts $0 \leqslant z \leqslant z_f$. For this purpose we will group together the traditional classes (spiral, elliptical, etc.) under the single heading of quiescent or *normal galaxies*. At higher redshifts ($z \gtrsim 1$), we will allow a second class of objects to play a role: the active or *starburst galaxies*. Whereas normal galaxies tend to be comprised of older, redder stellar populations, starburst galaxies are dominated by newly-forming stars whose energy output peaks in the ultraviolet (although much of this is absorbed by dust grains and subsequently reradiated in the infrared). One signature of the starburst type is thus a decrease in $F(\lambda)$ as a function of $\lambda$ over NUV and optical wavelengths, while normal types show an increase [76]. Starburst galaxies also tend to be brighter, reaching bolometric luminosities as high as $10^{12} - 10^{13} L_{\odot}$, versus $10^{10} - 10^{11} L_{\odot}$ for normal types.

There are two ways to obtain SEDs for these objects: by reconstruction from observational data, or as output from theoretical models of galaxy evolution. The former approach has had some success, but becomes increasingly difficult at short wavelengths, so that results have typically been

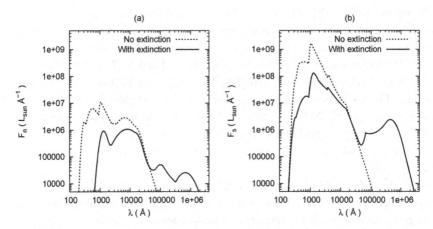

Fig. 3.5 Typical galaxy SEDs for (a) normal and (b) starburst type galaxies with and without extinction by dust. These figures are adapted from Figs. 9 and 10 of Ref. [77]. For definiteness we have normalized (over $100 - 3 \times 10^4$ Å) such that $L_n = 1 \times 10^{10} h_0^{-2} L_\odot$ and $L_s = 2 \times 10^{11} h_0^{-2} L_\odot$ with $h_0 = 0.75$. (These values are consistent with what we will later call "model 0" for a comoving galaxy number density of $n_0 = 0.010 h_0^3$ Mpc$^{-3}$.)

restricted to $\lambda \gtrsim 1000$Å [76]. This represents a serious limitation if we want to integrate out to redshifts $z_f \sim 6$ (say), since it means that our results are only strictly reliable down to $\lambda_0 = \lambda(1 + z_f) \sim 7000$Å. In order to integrate out to $z_f \sim 6$ and still go down as far as the NUV ($\lambda_0 \sim 2000$Å), we require SEDs which are good to $\lambda \sim 300$Å in the galaxy rest-frame. For this purpose we will make use of theoretical galaxy-evolution models, which have advanced to the point where they cover the entire spectrum from the far ultraviolet to radio wavelengths. This broad range of wavelengths involves diverse physical processes such as star formation, chemical evolution, and (of special importance here) dust absorption of ultraviolet light and re-emission in the infrared. Typical normal and starburst galaxy SEDs based on such models are now available down to $\sim 100$Å [77]. These functions, displayed in Fig. 3.5, will constitute our normal and starburst galaxy SEDs, $F_n(\lambda)$ and $F_s(\lambda)$.

Fig. 3.5 shows the expected increase in $F_n(\lambda)$ with $\lambda$ at NUV wavelengths ($\sim 2000$Å) for normal galaxies, as well as the corresponding decrease for starbursts. What is most striking about both templates, however, is their overall multi-peaked structure. These objects are far from pure blackbodies, and the primary reason for this is *dust*. This effectively removes light from the shortest-wavelength peaks (which are due mostly to star formation), and transfers it to the longer-wavelength ones. The dashed lines in Fig. 3.5 show what the SEDs would look like if this dust reprocessing were ignored. The main difference between normal and starburst types lies in the relative importance of this process. Normal galaxies emit as little as 30% of their bolometric intensity in the infrared, while the equivalent fraction for the largest starburst galaxies can reach 99%. Such variations can be incorporated by modifying input parameters such as star formation timescale and gas density, leading to spectra which are broadly similar in shape to those in Fig. 3.5 but differ in normalization and "tilt" toward longer wavelengths. The results have been successfully matched to a wide range of real galaxy spectra [77].

## 3.7 The light of the night sky

We proceed to calculate the spectral EBL intensity using $F_n(\lambda)$ and $F_s(\lambda)$, with the characteristic luminosities of these two types found as usual by normalization, $\int F_n(\lambda)d\lambda = L_n$ and $\int F_s(\lambda)d\lambda = L_s$. Let us assume that the comoving luminosity density of the Universe at any redshift $z$ is a combination of normal and starburst components

$$\mathcal{L}(z) = n_n(z)L_n + n_s(z)L_s , \qquad (3.29)$$

where comoving number densities are

$$n_n(z) \equiv [1 - f(z)]\, n(z) \qquad n_s(z) \equiv f(z)\, n(z) . \qquad (3.30)$$

In other words, we will account for evolution in $\mathcal{L}(z)$ solely in terms of a changing *starburst fraction* $f(z)$, and a single comoving number density $n(z)$ as before. $L_n$ and $L_s$ are awkward to work with for dimensional reasons, and we will find it more convenient to specify the SED instead by two dimensionless parameters, the local starburst fraction $f_0$ and luminosity ratio $\ell_0$:

$$f_0 \equiv f(0) \qquad \ell_0 \equiv L_s/L_n . \qquad (3.31)$$

Observations indicate that $f_0 \approx 0.05$ in the local population [76], and the SEDs shown in Fig. 3.5 have been fitted to a range of normal and starburst galaxies with $40 \lesssim \ell_0 \lesssim 890$ [77]. We will allow these two parameters to vary in the ranges $0.01 \leqslant f_0 \leqslant 0.1$ and $10 \leqslant \ell_0 \leqslant 1000$. This, in combination with our "strong" and "weak" limits on luminosity-density evolution, gives us the flexibility to obtain upper and lower bounds on EBL intensity.

The functions $n(z)$ and $f(z)$ can now be fixed by equating $\mathcal{L}(z)$ as defined by (3.29) to the comoving luminosity-density curves inferred from HDF data (Fig. 3.1), and requiring that $f \to 1$ at peak luminosity (i.e. assuming that the galaxy population is entirely starburst-dominated at the redshift $z_p$ of peak luminosity). These conditions are not difficult to set up. One finds that modest number-density evolution is required in general, if $f(z)$ is not to over- or under-shoot unity at $z_p$. We follow [79] and parametrize this with the function $n(z) = n_0(1 + z)^\eta$ for $z \leqslant z_p$. Here $\eta$ is often termed the *merger parameter* since a value of $\eta > 0$ would imply that the comoving number density of galaxies decreases with time.

Pulling these requirements together, one obtains a model with

$$
f(z) = \begin{cases} \left(\dfrac{1}{\ell_0 - 1}\right)[\ell_0(1 + z)^{-\eta}\mathcal{N}(z) - 1] & (z \leqslant z_p) \\ 1 & (z > z_p) \end{cases}
$$

$$
n(z) = n_0 \times \begin{cases} (1 + z)^\eta & (z \leqslant z_p) \\ \mathcal{N}(z) & (z > z_p) \end{cases}. \tag{3.32}
$$

Here $\mathcal{N}(z) \equiv [1/\ell_0 + (1 - 1/\ell_0)f_0]\,\tilde{\mathcal{L}}(z)$ and $\eta = \ln[\mathcal{N}(z_p)]/\ln(1 + z_p)$. The evolution of $f(z)$, $n_n(z)$ and $n_s(z)$ is plotted in Fig. 3.6 for five models: a best-fit Model 0, corresponding to the moderate evolution curve in Fig. 3.1 with $f_0 = 0.05$ and $\ell_0 = 20$, and four other models chosen to produce the widest possible spread in EBL intensities across the optical band. Models 1 and 2 are the *most* starburst-dominated, with initial starburst fraction and luminosity ratio at their upper limits ($f_0 = 0.1$ and $\ell_0 = 1000$). Models 3 and 4 are the *least* starburst-dominated, with the same quantities at their lower limits ($f_0 = 0.01$ and $\ell_0 = 10$). Luminosity density evolution is set to "weak" in the odd-numbered Models 1 and 3, and "strong" in the even-numbered Models 2 and 4. We find merger parameters $\eta$ between $+0.4, 0.5$ in the strong-evolution Models 2 and 4, and $-0.5, -0.4$ in the weak-evolution Models 1 and 3, while $\eta = 0$ for Model 0. These are well within the normal range [80].

The information contained in Fig. 3.6 can be summarized in words as follows: starburst galaxies formed near $z_f \sim 4$ and increased in comoving number density until $z_p \sim 2.5$ (the redshift of peak comoving luminosity density in Fig. 3.1). They then gave way to a steadily growing population of fainter normal galaxies which began to dominate between $1 \lesssim z \lesssim 2$ (depending on the model) and now make up 90–99% of the total galaxy population at $z = 0$. This scenario is in good agreement with others that have been constructed to explain the observed faint blue excess in galaxy number counts [78].

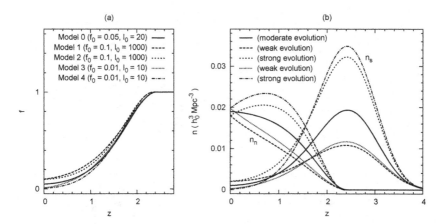

Fig. 3.6 Evolution of (a) starburst fraction $f(z)$ and (b) comoving normal and starburst galaxy number densities $n_n(z)$ and $n_s(z)$, where total comoving luminosity density $\mathcal{L}(z) = n_n(z)L_n + n_s(z)L_s$ is matched to the "moderate," "weak" and "strong" evolution curves in Fig. 3.1. Each model has different values of the two adjustable parameters $f_0 \equiv f(0)$ and $\ell_0 \equiv L_s/L_n$.

The total spectral EBL intensity can now be obtained by substituting the SEDs $(F_n, F_s)$ and comoving number densities (3.30) into Eq. (3.6). We write the results in the form $I_\lambda(\lambda_0) = I_\lambda^n(\lambda_0) + I_\lambda^s(\lambda_0)$ where:

$$I_\lambda^n(\lambda_0) = I_{ns} \int_0^{z_f} \tilde{n}(z)[1 - f(z)] F_n\left(\frac{\lambda_0}{1+z}\right) \frac{dz}{(1+z)^3 \tilde{H}(z)} ,$$

$$I_\lambda^s(\lambda_0) = I_{ns} \int_0^{z_f} \tilde{n}(z)f(z) F_s\left(\frac{\lambda_0}{1+z}\right) \frac{dz}{(1+z)^3 \tilde{H}(z)} . \qquad (3.33)$$

Here $I_\lambda^n$ and $I_\lambda^s$ represent contributions from normal and starburst galaxies respectively and $\tilde{n}(z) \equiv n(z)/n_0$ is the *relative comoving number density*. The dimensional content of both integrals has been pulled into a prefactor

$$I_{ns} = I_{ns}(\lambda_0) = \frac{\mathcal{L}_0}{4\pi h H_0}\left(\frac{\lambda_0}{\text{Å}}\right) = 970 \text{ CUs}\left(\frac{\lambda_0}{\text{Å}}\right) . \qquad (3.34)$$

This is independent of $h_0$, as before, because the factor of $h_0$ in $\mathcal{L}_0$ cancels out the one in $H_0$. The quantity $\mathcal{L}_0$ appears here when we normalize the galaxy SEDs $F_n(\lambda)$ and $F_s(\lambda)$ to the observed comoving luminosity density of the Universe. To see this, note that Eq. (3.29) reads $\mathcal{L}_0 = n_0 L_n[1 + (\ell_0 - 1)f_0]$ at $z = 0$. Since $\mathcal{L}_0 \equiv n_0 L_0$, it follows that $L_n = L_0/[1 + (\ell_0 - 1)f_0]$ and $L_s = L_0 \ell_0/[1 + (\ell_0 - 1)f_0]$. Thus a factor of $L_0$ can be divided out of the functions $F_n$ and $F_s$ and put directly into Eq. (3.33) as required.

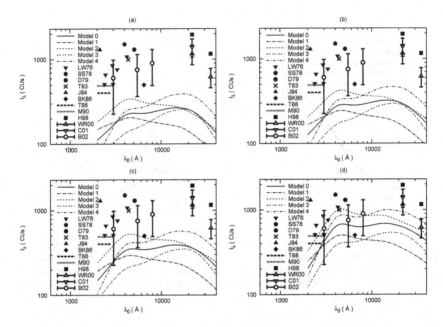

Fig. 3.7   The spectral EBL intensity of a combined population of normal and starburst galaxies, with SEDs as shown in Fig. 3.5. The evolving number densities are such as to reproduce the total comoving luminosity density seen in the HDF (Fig. 3.1). Results are shown for the (a) EdS, (b) OCDM, (c) ΛCDM and (d) ΛBDM cosmologies. Also shown are observational upper limits (solid symbols and heavy lines) and reported detections (open symbols).

The spectral intensity (3.33) is plotted in Fig. 3.7, where we have set $z_f = 6$ as usual. (Results are insensitive to this choice, dropping by less than 5% for $z_f = 3$.) These plots show that the most starburst-dominated models (1 and 2) produce the bluest EBL spectra, as might be expected. For these two models, EBL contributions from normal galaxies remain well below those from starbursts at all wavelengths, so that the bump in the observed spectrum at $\lambda_0 \sim 4000$Å is essentially an echo of the peak at $\sim 1100$Å in the starburst SED (Fig. 3.5), redshifted by a factor $(1 + z_p)$ from the epoch $z_p \approx 2.5$ of maximum comoving luminosity density. By contrast, in the least starburst-dominated models (3 and 4), EBL contributions from normal galaxies catch up to and exceed those from starbursts at $\lambda_0 \gtrsim 10,000$Å, giving rise to the bump seen at $\lambda_0 \sim 20,000$Å in these models. Absolute EBL intensities are highest in the strong-evolution models (2 and 4) and lowest in the weak-evolution models (1 and 3). We emphasize that the *total* amount of light in the EBL is determined by the choice of luminosity density profile (for a given cosmological model). The choice of SED merely shifts this light from one part of the spectrum to another. Within the context of the simple two-component model described above, and the constraints imposed on luminosity density by the HDF data (Sec. 3.2), the curves in Fig. 3.7 represent upper and lower limits on the spectral intensity of the EBL at near-optical wavelengths.

These results are more broadly distributed in wavelength, and fainter in peak intensity, than those obtained earlier using single-component Gaussian and blackbody spectra. In fact these predicted EBL intensities are now significantly fainter than those actually observed in all but the most vacuum-dominated cosmology, ΛBDM (d). This is so even for the models with the strongest luminosity density evolution (models 2 and 4). In the case of the EdS cosmology (a), the gap is nearly an order of magnitude, in agreement with earlier analyses [74]. Similar conclusions have been drawn from an analysis of Subaru Deep Field data [81], with a suggestion that the shortfall could be made up by a diffuse, previously undetected component of background radiation not associated with galaxies. Others have argued that existing galaxy populations are enough to explain the data if different assumptions are made about their SEDs [82], or if allowance is made for faint, low surface-brightness galaxies (LSBs) below the detection limit of existing surveys [83].

## 3.8 R.I.P. Olbers' paradox

Having obtained quantitative estimates of the spectral EBL intensity which are in reasonable agreement with observation, we return to the question posed in Sec. 2.4: Why, precisely, is the sky dark at night? By "dark" we now mean specifically dark at *near-optical wavelengths*. We can provide a quantitative answer to this question by using a spectral version of our previous bolometric argument. That is, we compute the EBL intensity $I_{\lambda,\text{stat}}$ in model universes which are equivalent to expanding ones in every way *except* expansion, and then take the ratio $I_\lambda / I_{\lambda,\text{stat}}$. If this is of order unity, then expansion plays a minor role and the darkness of the optical sky (like the bolometric one) must be attributed mainly to the fact that the Universe is too young to have filled up with light. If $I_\lambda / I_{\lambda,\text{stat}} \ll 1$, on the other hand, then we would have a situation qualitatively different from the bolometric one, and expansion would play a crucial role in the resolution to Olbers' paradox.

The spectral EBL intensity for the equivalent static model is obtained by putting the functions $\tilde{n}(z), f(z), F_n(\lambda), F_s(\lambda)$ and $\tilde{H}(z)$ into (3.8) rather than (3.6). This results in $I_{\lambda,\text{stat}}(\lambda_0) = I_{\lambda,\text{stat}}^n(\lambda_0) + I_{\lambda,\text{stat}}^s(\lambda_0)$, where normal and starburst contributions are given by

$$I_{\lambda,\text{stat}}^n(\lambda_0) = I_{ns} F_n(\lambda_0) \int_0^{z_f} \frac{\tilde{n}(z)[1 - f(z)] \, dz}{(1 + z)\tilde{H}(z)}$$

$$I_{\lambda,\text{stat}}^s(\lambda_0) = I_{ns} F_s(\lambda_0) \int_0^{z_f} \frac{\tilde{n}(z)f(z) \, dz}{(1 + z)\tilde{H}(z)} \,. \tag{3.35}$$

Despite a superficial resemblance to their counterparts (3.33) for expanding models, these are vastly different expressions. The SEDs $F_n(\lambda_0)$ and $F_s(\lambda_0)$ no longer depend on $z$ and have been pulled out of the integrals. The quantity $I_{\lambda,\text{stat}}(\lambda_0)$ is effectively a *weighted mean* of the SEDs $F_n(\lambda_0)$ and $F_s(\lambda_0)$. The weighting factors are related to the age of the galaxies, $\int_0^{z_f} dz/(1 + z)\tilde{H}(z)$, but modified by factors of $n_n(z)$ and $n_s(z)$ under the integral. This latter modification is important because it prevents the integrals from increasing without limit as $z_f$ becomes arbitrarily large, a problem that would otherwise introduce considerable uncertainty into any attempt to put bounds on the ratio $I_{\lambda,\text{stat}}/I_\lambda$ [37].

The ratio of $I_\lambda/I_{\lambda,\text{stat}}$ is plotted over the waveband 2000-25,000Å in Fig. 3.8, where we have set $z_f = 6$. (Results are insensitive to this choice, and also independent of uncertainty in constants such as $\mathcal{L}_0$ since these are common to both $I_\lambda$ and $I_{\lambda,\text{stat}}$.) Several features in this figure deserve comment. First, the average value of $I_\lambda/I_{\lambda,\text{stat}}$ across the spectrum is about 0.6, consistent with bolometric expectations (Chap. 2). Second, the diagonal, bottom-left to top-right orientation arises largely because $I_\lambda(\lambda_0)$ drops off at short wavelengths, while $I_{\lambda,\text{stat}}(\lambda_0)$ does so at long ones. The reason why $I_\lambda(\lambda_0)$ drops off at short wavelengths is that ultraviolet light reaches us only from the nearest galaxies; anything from more distant ones is redshifted into the optical. The reason why $I_{\lambda,\text{stat}}(\lambda_0)$ drops off at long wavelengths is because it is a weighted mixture of the galaxy SEDs, and drops off at exactly the same place that they do: $\lambda_0 \sim 3 \times 10^4$Å. In fact, the weighting is heavily tilted toward the dominant starburst component, so that the two sharp bends apparent in Fig. 3.8 are essentially (inverted) reflections of features in $F_s(\lambda_0)$; namely, the small bump at $\lambda_0 \sim 4000$Å and the shoulder at $\lambda_0 \sim 11,000$Å (Fig. 3.5).

Finally, the numbers: Fig. 3.8 shows that the ratio of $I_\lambda/I_{\lambda,\text{stat}}$ is remarkably consistent across the B-band (4000-5000Å) in all four cosmological models, varying from a high of $0.46 \pm 0.10$ in the EdS model to a low of

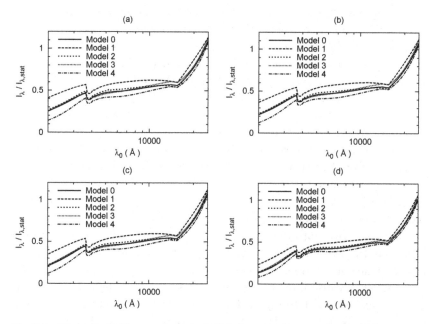

Fig. 3.8    The ratio $I_\lambda/I_{\lambda,\text{stat}}$ of spectral EBL intensity in expanding models to that in equivalent static models, for the (a) EdS, (b) OCDM, (c) $\Lambda$CDM and (d) $\Lambda$BDM models. The fact that this ratio lies between 0.3 and 0.6 in the B-band (4000-5000Å) tells us that expansion reduces the intensity of the night sky at optical wavelengths by a factor of between two and three.

$0.39 \pm 0.08$ in the $\Lambda$BDM model. These numbers should be compared with the bolometric result of $Q/Q_{\text{stat}} \approx 0.6 \pm 0.1$ from Chap. 2. They tell us that expansion *does* play a greater role in determining B-band EBL intensity than it does across the spectrum as a whole—but not by much. If the universe were not expanding, the night sky at optical wavelengths would be between two times brighter (in the EdS model) and three times brighter (in the $\Lambda$BDM model). These results depend modestly on the makeup of the evolving galaxy population, and Fig. 3.8 shows that $I_\lambda/I_{\lambda,\text{stat}}$ in every case is highest for the weak-evolution model 1, and lowest for the strong-evolution model 4. This is as we would expect, based on our discussion at the beginning of this section: models with the strongest evolution effectively "concentrate" their light production over the shortest possible interval in time, so that the importance of the lifetime factor drops relative to that of expansion. Our numerical results, however, prove that this effect cannot qualitatively alter the resolution of Olbers' paradox. Whether expansion reduces the background intensity by a factor of two or three, its *order of magnitude* is still set by the lifetime of the Universe.

There is one factor which we have not considered in this section, and that is extinction by intergalactic dust and neutral hydrogen, both of which are strongly absorbing at ultraviolet wavelengths. The effect of this is primarily to remove ultraviolet light from high-redshift galaxies and transfer it to the infrared—light that would otherwise be redshifted into the optical and contribute to the EBL. The latter's intensity $I_\lambda(\lambda_0)$ therefore drops, and reductions can be expected over the B-band in particular. This effect, if significant, would further widen the gap between observed and predicted EBL intensities noted at the end of Sec. 3.6.

Absorption plays far less of a role in the corresponding static models, where there is no redshift. (Ultraviolet light is still absorbed, but the effect does not carry over into the optical). Therefore, the ratio $I_\lambda/I_{\lambda,\text{stat}}$ would be expected to drop in nearly direct proportion to the drop in $I_\lambda$. In this sense de Chéseaux and Olbers may have had part of the solution after all; not so much because intervening matter "absorbs" the light from distant sources, but because it *transfers it out of the optical*. The quantitative importance of this effect is difficult to assess because we have limited data on the character and distribution of dust beyond our own galaxy. We will find indications in Chap. 7 that the reduction in intensity could be significant, but only at the shortest wavelengths considered here ($\lambda_0 \approx 2000$Å), and only for the dust models whose properties are tuned to give the highest possible UV extinction while still remaining consistent with other constraints. Thus, if expansion produces at most a first-order correction to the intensity of the EBL, then the effects of absorption are at most of second order. The *zeroth*-order solution to Olbers' paradox, as demonstrated quantitatively above, remains the finite age of the Universe and the galaxies it contains. The optical sky, like the bolometric one, is dark at night primarily because it has not had enough time to fill up with light.

# Chapter 4

# Dark Cosmology

## 4.1 The four dark elements

Observations of the intensity of extragalactic background light effectively allow us to take a census of one component of the Universe: its luminous matter. In astronomy, where nearly everything we know comes to us in the form of light signals, one might be forgiven for thinking that luminous matter was the only kind that counted. This supposition, however, turns out to be spectacularly wrong. The density $\Omega_{lum}$ of luminous matter is now thought to comprise less than one percent of the total density $\Omega_{tot}$ of all forms of matter and energy put together. The remaining 99% or more consists of new forms of matter-energy, not detectable by conventional telescopes, whose existence is inferred solely from their gravitational influence on the matter we can see.

Since their existence was first suspected in the 1920s and 1930s by astronomers such as J.C. Kapteyn [84], J.H. Oort [85] and F. Zwicky [86], the identity of these new forms of matter-energy has become the greatest unsolved mystery in cosmology. Indirect evidence over the past decade has increasingly suggested that there are in fact *four distinct categories* of dark matter and energy, three of which imply new physics beyond the existing standard model of particle interactions. This is an extraordinary claim, and one whose supporting evidence deserves to be scrutinized carefully. We devote Chap. 4 to a critical review of this evidence, beginning here with a brief overview of the current situation and followed by a closer look at the arguments for all four parts of nature's "dark side."

At least some of the dark matter, such as that contained in planets and "failed stars" too dim to see, must be composed of ordinary atoms and molecules. The same applies to dark gas and dust (although these can sometimes be seen in absorption, if not emission). Such contributions comprise *baryonic dark matter* (BDM), which combined together with luminous matter gives a total baryonic matter density of $\Omega_{bar} \equiv \Omega_{lum} + \Omega_{bdm}$. If our understanding of big-bang theory and the formation of the light elements is correct, then we will see that $\Omega_{bar}$ cannot represent more than 5% of the critical density.

Besides the dark baryons, it now appears that three other varieties of dark matter play a role. The first of these is *cold dark matter* (CDM), the existence of which has been inferred from the behavior of visible matter on scales larger than the solar system (e.g., galaxies and clusters of galaxies). CDM is thought to consist of particles (sometimes referred to as "exotic" dark-matter particles) whose interactions with ordinary matter are so weak that they are seen primarily via their gravitational influence. While they have not been detected (and are indeed hard to detect by definition), such particles are predicted in plausible extensions of the standard model. The overall CDM density $\Omega_{cdm}$ is believed by many cosmologists to exceed that of the baryons ($\Omega_{bar}$) by at least an order of magnitude.

Another piece of the puzzle is provided by *neutrinos*, particles whose existence is unquestioned but whose collective density ($\Omega_\nu$) depends on their rest masses, which are not precisely known. If neutrinos are massless, or nearly so, then they remain relativistic throughout the history of the Universe and behave for dynamical purposes like photons. In this case neutrino contributions combine with those of photons ($\Omega_\gamma$) to give the present radiation-energy density as $\Omega_{r,0} = \Omega_\nu + \Omega_\gamma$. This is known to be very small. If on the other hand neutrinos are sufficiently massive, then they are no longer relativistic on average, and belong together with baryonic and cold dark matter under the category of pressureless matter, with present density $\Omega_{m,0} = \Omega_{bar} + \Omega_{cdm} + \Omega_\nu$. These neutrinos could play a significant dynamical role, especially in the formation of large-scale structures in the early Universe, where they are sometimes known as hot dark matter (HDM). Recent experimental evidence suggests that neutrinos do contribute to $\Omega_{m,0}$ but at levels below those of the baryons.

Influential only over the largest scales—those of the cosmological horizon itself—is the final component of the unseen Universe: *dark energy*. Its many alternative names (the zero-point field, vacuum energy, quintessence and the cosmological constant $\Lambda$) testify to the fact that there is currently no consensus as to where dark energy originates, or how to calculate its energy density ($\Omega_\Lambda$) from first principles. Existing theoretical estimates of this latter quantity range over many orders of magnitude, prompting some

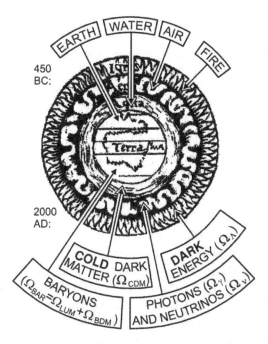

Fig. 4.1 Top: earth, water, air and fire, the four elements of ancient cosmology (attributed to the Greek philosopher Empedocles). Bottom: their modern counterparts (figure taken from the review in Ref. [87]).

cosmologists until very recently to disregard it altogether. Observations of distant supernovae, however, increasingly imply that dark energy is not only real but that its present energy density $(\Omega_{\Lambda,0})$ exceeds that of all other forms of matter $(\Omega_{m,0})$ and radiation $(\Omega_{r,0})$ put together.

The Universe described above hardly resembles the one we see. It is composed to a first approximation of invisible dark energy whose physical origin remains obscure. Most of what remains is in the form of CDM particles, whose exotic nature is also not yet understood. Close inspection is needed to make out the further contribution of neutrinos, although this too is nonzero. And baryons, the stuff of which we are made, are little more than a cosmic afterthought. This picture entails a shift in the way we see the Universe that is profound enough to be called a Copernican counter-revolution. For while our location in space may be undistinguished, our *composition*, and that of our local environment, is truly atypical. The "four elements" of modern cosmology are shown juxtaposed with their ancient counterparts in Fig. 4.1.

## 4.2 Baryons

Let us now go over the evidence for these four species of dark matter more carefully, beginning with the baryons. The total present density of *luminous* baryonic matter can be inferred from the observed luminosity density of the Universe, if various reasonable assumptions are made about the fraction of galaxies of different morphological type, their ratios of disk-type to bulge-type stars, and so on. A typical result is [40]:

$$\Omega_{lum} = (0.0027 \pm 0.0014)h_0^{-1} \ . \tag{4.1}$$

Here $h_0$ is as usual the value of Hubble's constant expressed in units of 100 km s$^{-1}$ Mpc$^{-1}$. While this parameter (and hence the experimental uncertainty in $H_0$) factored out of the EBL intensities in Chaps. 2 and 3, it must be squarely faced where densities are concerned. We therefore digress briefly to discuss the observational status of $h_0$.

Using various relative-distance methods, all calibrated against the distance to Cepheid variables in the Large Magellanic Cloud (LMC), the Hubble Key Project (HKP) has determined that $h_0 = 0.72 \pm 0.08$ [88]. Independent "absolute" methods (e.g., time delays in gravitational lenses, the Sunyaev-Zeldovich effect and the Baade-Wesselink method applied to supernovae) have higher uncertainties but are roughly consistent with this, giving $h_0 \approx 0.55 - 0.74$ [31]. This level of agreement is a great improvement over the factor-two discrepancies of previous decades.

There are signs, however, that we are still some way from "precision" values with uncertainties of less than ten percent. A recalibrated LMC Cepheid period-luminosity relation based on a much larger sample (from the OGLE microlensing survey) leads to considerably higher values, namely $h_0 = 0.85 \pm 0.05$ [89]. A purely geometric technique, based on the use of long-baseline radio interferometry to measure the transverse velocity of water masers [90], also implies that the traditional calibration is off, raising all Cepheid-based estimates by $12 \pm 9\%$ [91]. This would boost the HKP value to $h_0 = 0.81 \pm 0.09$. There is some independent support for such a recalibration in observations of "red-clump stars" [92] and eclipsing binaries [93] in the LMC.

On this subject, history encourages caution. Where it is necessary to specify the value of $h_0$ in this book, we will adopt:

$$h_0 = 0.75 \pm 0.15 \ . \tag{4.2}$$

Values at the edges of this range can discriminate powerfully between different cosmological models. This is largely a function of their *ages*, which

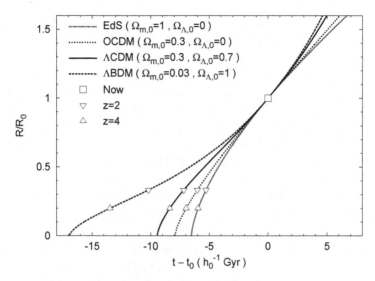

Fig. 4.2   Evolution of the cosmological scale factor $\tilde{R}(t) \equiv R(t)/R_0$ as a function of time (in Hubble times) for the cosmological test models introduced in Table 3.1 (Chap. 3). Triangles indicate the range $2 \leqslant z_f \leqslant 4$ where the bulk of galaxy formation may have taken place.

can be computed by integrating (2.34) or (for 3D-flat models) directly from Eq. (2.54). Alternatively, one can integrate the Friedmann-Lemaître equation (2.31) numerically backward in time. Since this equation defines the expansion rate $H \equiv \dot{R}/R$, its integral gives the scale factor $R(t)$. We plot the results in Fig. 4.2 for the four cosmological "test models" in Table 3.1 (EdS, OCDM, ΛCDM, and ΛBDM). These are seen to have ages of $7h_0^{-1}$, $8h_0^{-1}$, $10h_0^{-1}$ and $17h_0^{-1}$ Gyr respectively. A firm lower limit of 11 Gyr can be set on the age of the Universe by means of certain metal-poor halo stars whose ratios of radioactive $^{232}$Th and $^{238}$U imply that they formed between $14.1 \pm 2.5$ Gyr [94] and $15.5 \pm 3.2$ Gyr ago [95]. If $h_0$ lies at the upper end of the above range ($h_0 = 0.9$), then the EdS and OCDM models would be ruled out on the basis that they are not old enough to contain these stars (this is known as the *age crisis* in low-$\Omega_{\Lambda,0}$ models). With $h_0$ at the bottom of the range ($h_0 = 0.6$), however, only EdS comes close to being excluded. The EdS model thus defines one edge of the spectrum of observationally-viable models.

The ΛBDM model faces the opposite problem: Fig. 4.2 shows that its age is $17h_0^{-1}$ Gyr, or as high as 28 Gyr (if $h_0 = 0.6$). The latter number in particular is well beyond the age of anything seen in our Galaxy. Of course, upper limits on the age of the Universe are not as secure as lower ones. But following Copernican reasoning, we do not expect to live in a galaxy which is unusually young. To estimate the age of a "typical" galaxy, we recall from Sec. 3.2 that most galaxies appear to have formed at redshifts $2 \lesssim z_f \lesssim 4$. The corresponding range of scale factors, from Eq. (2.11), is $0.33 \gtrsim R/R_0 \gtrsim 0.2$. In the $\Lambda$BDM model, Fig. 4.2 shows that $R(t)/R_0$ does not reach these values until $(5 \pm 2) h_0^{-1}$ Gyr after the big bang. Thus galaxies would have an age of about $(12 \pm 2) h_0^{-1}$ Gyr in the $\Lambda$BDM model, and not more than 21 Gyr in any case. This is close to upper limits which have been set on the age of the Universe in models of this type, $t_0 < 24 \pm 2$ Gyr [96]. Thus the $\Lambda$BDM model, or something close to it, probably defines a position opposite that of EdS on the spectrum of observationally-viable models.

For the other three models, Fig. 4.2 shows that galaxy formation must be accomplished within less than 2 Gyr after the big bang. The reason this is able to occur so quickly is that these models all contain significant amounts of CDM, which decouples from the primordial plasma before the baryons and prepares potential wells for the baryons to fall into. This, as we will see, is one of the main motivations for CDM.

Returning now to the density of luminous matter, we find with our values (4.2) for $h_0$ that Eq. (4.1) gives $\Omega_{\text{lum}} = 0.0036 \pm 0.0020$. This is the basis for our statement (Sec. 4.1) that the *visible* components of the Universe account for less than about 1% of its density.

What about dark baryons? The theory of primordial big-bang nucleosynthesis provides us with an independent method for determining the density of *total* baryonic matter in the Universe, based on the assumption that the light elements we see today were forged in the furnace of the hot big bang. Results using different light elements are roughly consistent, which is impressive in itself. The primordial abundances of $^4$He (by mass) and $^7$Li (relative to H) imply a baryon density of $\Omega_{\text{bar}} = (0.010 \pm 0.004)h_0^{-2}$ [97]. By contrast, measurements based exclusively on the primordial D/H abundance give a higher value with lower uncertainty: $\Omega_{\text{bar}} = (0.019 \pm 0.002)h_0^{-2}$ [98]. Since it appears premature at present to exclude either of these results, we choose an intermediate value of $\Omega_{\text{bar}} = (0.016 \pm 0.005)h_0^{-2}$. Combining this with our range of values (4.2) for $h_0$, we conclude that

$$\Omega_{\text{bar}} = 0.028 \pm 0.012 . \tag{4.3}$$

This result provides the rationale for our choice of $\Omega_{m,0} = 0.03$ in the $\Lambda$BDM model, and also agrees with independent estimates obtained by adding up individual mass contributions from all known repositories of baryonic matter via their estimated mass-to-light ratios [40]. Eq. (4.3) implies that *all* the atoms and molecules in existence make up less than 5% of the critical density.

The vast majority of these baryons, moreover, are invisible. Using Eqs. (4.1) and (4.2) together with the above-mentioned value of $\Omega_{bar}h_0^2$, we infer a baryonic dark matter fraction $\Omega_{bdm}/\Omega_{bar} = 1 - \Omega_{lum}/\Omega_{bar} = (87 \pm 8)\%$. Where do these dark baryons reside? One possibility is that they are smoothly distributed in a gaseous intergalactic medium, which would have to be strongly ionized in order to explain why it has not left a more obvious absorption signature in the light from distant quasars. Observations using OVI absorption lines as a tracer of ionization suggest that the contribution of such material to $\Omega_{bar}$ is at least $0.003h_0^{-1}$ [99], comparable to $\Omega_{lum}$. Simulations are able to reproduce many observed features of the "forest" of Lyman-$\alpha$ (Ly$\alpha$) absorbers with as much as $80 - 90\%$ of the baryons in this form [100].

Dark baryonic matter could also be bound up in clumps of matter such as substellar objects (jupiters, brown dwarfs) or stellar remnants (white, red and black dwarfs, neutron stars, black holes). Substellar objects are not likely to make a large contribution, given their small masses. Black holes are limited in the opposite sense: they cannot be more massive than about $10^5 M_\odot$ since this would lead to dramatic tidal disruptions and lensing effects which are not seen [101]. The baryonic dark-matter clumps of most interest are therefore ones whose mass is within a few orders of magnitude of $M_\odot$. Gravitational microlensing constraints based on quasar variability do not seriously limit such objects at present, setting an upper bound of 0.1 (well above $\Omega_{bar}$) on their combined contributions to $\Omega_{m,0}$ in an EdS Universe [102].

The existence of at least one class of compact dark objects, the *massive compact halo objects* (MACHOs), has been confirmed within our own galactic halo by the MACHO microlensing survey of LMC stars [103]. The inferred lensing masses lie in the range $(0.15 - 0.9)M_\odot$ and would account for between 8% and 50% of the high rotation velocities seen in the outer parts of the Milky Way, depending on a choice of halo model. If the halo is spherical, isothermal and isotropic, then at most 25% of its mass can be ascribed to MACHOs, according to a complementary survey (EROS) of microlensing in the direction of the Small Magellanic Cloud (SMC) [104]. The identity of the lensing bodies discovered in these surveys has been hotly debated. White dwarfs are unlikely candidates, since we see no telltale metal-rich ejecta from their massive progenitors [105]. Low-mass red dwarfs would have to be older and/or fainter than usually assumed, based on the numbers seen so far [106]. Degenerate "beige dwarfs" that could form above the theoretical hydrogen-burning mass limit of $0.08M_\odot$ without fusing have been proposed as an alternative [107], but it appears that such stars would form far too slowly to be important [108].

## 4.3   Dark matter

The introduction of a second species of unseen dark matter into the Universe has been justified on three main grounds: (1) a range of observational arguments imply that the total density parameter of gravitating matter exceeds that provided by baryons and bound neutrinos; (2) our current understanding of the growth of large-scale structure (LSS) requires the process to be helped along by large quantities of non-relativistic, weakly interacting matter in the early Universe, creating the potential wells for infalling baryons; and (3) theoretical physics supplies several plausible (albeit still undetected) candidate CDM particles with the right properties.

Since our ideas on structure formation may change, and the candidate particles may not materialize, the case for cold dark matter turns at present on the observational arguments. At one time, these were compatible with $\Omega_{cdm} \approx 1$, raising hopes that CDM would resolve two of the biggest challenges in cosmology at a single stroke: accounting for LSS formation *and* providing all the dark matter necessary to make $\Omega_{m,0} = 1$, vindicating the EdS model (and with it, the simplest models of inflation). Observations, however, no longer support values of $\Omega_{m,0}$ this high, and independent evidence now points to the existence of at least two other forms of matter-energy beyond the standard model (neutrinos and dark energy). The CDM hypothesis is therefore no longer as compelling as it once was. With this in mind we will pay special attention to the observational arguments in this section. The *lower limit* on $\Omega_{m,0}$ is crucial: only if $\Omega_{m,0} > \Omega_{bar} + \Omega_\nu$ do we require $\Omega_{cdm} > 0$.

The arguments can be broken into two classes: those which are purely empirical, and those which assume in addition the validity of the *gravitational instability* (GI) theory of structure formation. Let us begin with the empirical arguments. The first has been mentioned already in Sec. 4.2: the spiral galaxy rotation curve. If the MACHO and EROS results are taken at face value, and if the Milky Way is typical, then compact objects make up less than 50% of the mass of the halos of spiral galaxies. If, as has been argued [109], the remaining halo mass cannot be attributed to baryonic matter in known forms such as dust, rocks, planets, gas, or hydrogen snowballs, then a more exotic form of dark matter is required.

The total mass of dark matter in galaxies, however, is limited. The easiest way to see this is to compare the *mass-to-light ratio* $(M/L)$ of our own Galaxy to that of the Universe as a whole. If the latter has the critical density, then its $M/L$-ratio is just the ratio of the critical density to its luminosity density: $(M/L)_{crit,0} = \rho_{crit,0}/\mathcal{L}_0 = (1040 \pm 230)M_\odot/L_\odot$, where we have used (2.18) for $\mathcal{L}_0$, (2.22) for $\rho_{crit,0}$ and (4.2) for $h_0$. The corresponding value for the Milky Way is $(M/L)_{mw} = (21 \pm 7)M_\odot/L_\odot$, since the latter's luminosity is $L_{mw} = (2.3 \pm 0.6) \times 10^{10}L_\odot$ (in the B-band) and its total dynamical mass (including that of any unseen halo component) is $M_{mw} = (4.9 \pm 1.1) \times 10^{11}M_\odot$ inside 50 kpc, from the motions of Galactic satellites [110]. The ratio of $(M/L)_{mw}$ to $(M/L)_{crit,0}$ is thus less than 3%, and even if we multiply this by a factor of a few to account for possible halo mass outside 50 kpc, it is clear that galaxies like our own cannot make up more than 10% of the critical density.

Most of the mass of the Universe, in other words, is spread over scales larger than galaxies, and it is here that the arguments for CDM take on the most force. The most straightforward of these involve further applications of the mass-to-light ratio: one measures $M/L$ for a chosen region, corrects for the corresponding value in the "field," and divides by $(M/L)_{crit,0}$ to obtain $\Omega_{m,0}$. Much, however, depends on the choice of region. A widely respected application of this approach is that of the CNOC group [111], which uses rich clusters of galaxies. These systems sample large volumes of the early Universe, have dynamical masses which can be measured by three independent methods (the virial theorem, x-ray gas temperatures and gravitational lensing), and are subject to reasonably well-understood evolutionary effects. They are found to have $M/L \sim 200M_\odot/L_\odot$ on average, giving $\Omega_{m,0} = 0.19 \pm 0.06$ when $\Omega_{\Lambda,0} = 0$ [111]. This result scales as $(1 - 0.4\Omega_{\Lambda,0})$ [112], so that $\Omega_{m,0}$ drops to $0.11 \pm 0.04$, for example, in a model with $\Omega_{\Lambda,0} = 1$.

The weak link in this chain of inference is that rich clusters may not be characteristic of the Universe as a whole. Only about 10% of galaxies are found in such systems. If *individual* galaxies (like the Milky Way, with $M/L \approx 21M_\odot/L_\odot$) are substituted for clusters, then the inferred value of $\Omega_{m,0}$ drops by a factor of ten, approaching $\Omega_{bar}$ and removing the need for CDM. An effort to address the impact of scale on $M/L$ arguments has led to the conclusion that $\Omega_{m,0} = 0.16 \pm 0.05$ (for flat models) when regions of all scales are considered from individual galaxies to superclusters [113].

Another line of argument is based on the *cluster baryon fraction*, or ratio of baryonic-to-total mass $(M_{bar}/M_{tot})$ in galaxy clusters. Baryonic matter is defined as the sum of visible galaxies and hot gas (the mass of which can be inferred from x-ray temperature data). Total cluster mass is measured by the above-mentioned methods (virial theorem, x-ray temperature, or gravitational lensing). At sufficiently large radii, the cluster may be taken as representative of the Universe as a whole, so that $\Omega_{m,0} = \Omega_{bar}/(M_{bar}/M_{tot})$, where $\Omega_{bar}$ is fixed by big-bang nucleosynthesis (Sec. 4.2). Applied to various clusters, this procedure leads to $\Omega_{m,0} = 0.3 \pm 0.1$ [114]. This result is probably an upper limit, partly because baryon enrichment is more likely to take place inside the cluster than out, and partly because *dark* baryonic matter (such as MACHOs) is not taken into account; this would raise $M_{bar}$ and lower $\Omega_{m,0}$.

Other direct methods of constraining the value of $\Omega_{m,0}$ are rapidly becoming available, including those based on the evolution of galaxy cluster x-ray temperatures [115] and distortions in the images of distant galaxies due to weak gravitational lensing by intervening large-scale structures [116]. In combination with other evidence to be discussed shortly, especially that based on type Ia supernovae (SNIa) and fluctuations in the cosmic microwave background (CMB), these techniques show considerable promise for reducing the uncertainty in the matter density.

We move next to measurements of $\Omega_{m,0}$ based on the assumption that the growth of LSS proceeded via gravitational instability from a Gaussian spectrum of primordial density fluctuations (GI theory for short). These argmuents are circular in the sense that such a process could not have taken place as it did *unless* $\Omega_{m,0}$ is considerably larger than $\Omega_{bar}$. But inasmuch as GI theory is currently accepted as the most plausible framework for understanding cosmological structure formation on large scales (albeit with some difficulties on smaller scales [117]), this way of determining $\Omega_{m,0}$ should be taken seriously.

According to GI theory, LSS formation is more or less complete by $z \approx \Omega_{m,0}^{-1} - 1$ [49]. Therefore, one way to constrain $\Omega_{m,0}$ is to look for *number density evolution* in large-scale structures such as galaxy clusters. In a low-matter-density Universe, this would be relatively constant out to at least $z \sim 1$, whereas in a high-matter-density Universe one would expect the abundance of clusters to drop rapidly with $z$ because they are still in the process of forming. The fact that massive clusters are seen at redshifts as high as $z = 0.83$ has been used to infer that $\Omega_{m,0} = 0.17^{+0.14}_{-0.09}$ for $\Omega_{\Lambda,0} = 0$ models, and $\Omega_{m,0} = 0.22^{+0.13}_{-0.07}$ for flat ones [118].

Studies of the *power spectrum* $P(k)$ of the distribution of galaxies or other structures can be used in a similar way. In GI theory, structures of a given mass form by the collapse of large volumes in a low-matter-density Universe, or smaller volumes in a high-matter-density Universe. Thus $\Omega_{m,0}$ can be constrained by changes in $P(k)$ between one redshift and another. Comparison of the mass power spectrum of Ly$\alpha$ absorbers at $z \approx 2.5$ with that of local galaxy clusters at $z = 0$ has led to an estimate of $\Omega_{m,0} = 0.46^{+0.12}_{-0.10}$ for $\Omega_{\Lambda,0} = 0$ models [119]. This result goes as approximately $(1 - 0.4\Omega_{\Lambda,0})$, so that the central value of $\Omega_{m,0}$ drops to 0.34 in a flat model, and 0.28 if $\Omega_{\Lambda,0} = 1$. One can also constrain $\Omega_{m,0}$ from the local galaxy power spectrum alone, although this involves some assumptions about the extent to which "light traces mass" (i.e. to which visible galaxies trace the underlying density field). Results from the 2dF survey give $\Omega_{m,0} = 0.29^{+0.12}_{-0.11}$ assuming $h_0 = 0.7 \pm 0.1$ [120]. An early fit from the Sloan Digital Sky Survey (SDSS) is $\Omega_{m,0} = 0.19^{+0.19}_{-0.11}$ [121]. There are good prospects for reducing this uncertainty by combining data from galaxy surveys of this kind with CMB data [122], though such results must be interpreted with care at present [123].

A third group of measurements, and one which has traditionally yielded the highest estimates of $\Omega_{m,0}$, involves *galaxy peculiar velocities*. These are generated by the gravitational potential of locally over or under-dense regions relative to the mean matter density. The power spectra of the velocity and density distributions can be related to each other in the context of GI theory in a way which depends explicitly on $\Omega_{m,0}$. Tests of this kind probe relatively small volumes and are hence insensitive to $\Omega_{\Lambda,0}$, but they can depend significantly on $h_0$ as well as the spectral index $n$ of the density distribution. Where the latter is normalized to CMB fluctuations, results [124] take the form $\Omega_{m,0} h_0^{1.3} n^2 \approx 0.33 \pm 0.07$ or (taking $n = 1$ and using our values of $h_0$) $\Omega_{m,0} \approx 0.48 \pm 0.15$.

In summarizing these results, one is struck by the fact that arguments based on gravitational instability (GI) theory favor values of $\Omega_{m,0} \gtrsim 0.2$ and *higher*, whereas purely empirical arguments require $\Omega_{m,0} \lesssim 0.4$ and *lower*. The latter are in fact compatible in some cases with values of $\Omega_{m,0}$ as low as $\Omega_{\rm bar}$, raising the possibility that CDM might not in fact be necessary. The results from GI-based arguments, however, cannot be stretched this far. What is sometimes done is to "go down the middle" and combine the results of both kinds of argument into a single expression of the form $\Omega_{m,0} \approx 0.3 \pm 0.1$. Any such expression with $\Omega_{m,0} > 0.05$ constitutes a proof of the existence of CDM, since $\Omega_{\text{bar}} \leqslant 0.04$ from (4.3). (Neutrinos only strengthen this argument, as we note in Sec. 4.4.) A more conservative interpretation of the data, bearing in mind the full range of $\Omega_{m,0}$ values implied above ($\Omega_{\text{bar}} \lesssim \Omega_{m,0} \lesssim 0.6$), is

$$\Omega_{\text{cdm}} = 0.3 \pm 0.3 . \tag{4.4}$$

But it should be stressed that values of $\Omega_{\text{cdm}}$ at the bottom of this range carry with them the (uncomfortable) implication that the conventional picture of structure formation via gravitational instability is incomplete. Conversely, *if our current understanding of structure formation is correct, then CDM must exist* and $\Omega_{\text{cdm}} > 0$.

The question, of course, becomes moot if CDM is discovered in the laboratory. From a large field of theoretical particle candidates, two have emerged as frontrunners: axions and supersymmetric weakly-interacting massive particles (WIMPs). The plausibility of both candidates rests on three properties: they are (1) *weakly interacting* (i.e. "noticed" by ordinary matter primarily via their gravitational influence); (2) *cold* (i.e. non-relativistic in the early Universe, when structures began to form); and (3) *expected* on theoretical grounds to have a collective density within a few orders of magnitude of the critical one. We will return to these particles in Chaps. 6 and 8 respectively.

## 4.4 Neutrinos

Since neutrinos indisputably exist in great numbers, they have been leading dark-matter candidates for longer than axions or WIMPs. They gained prominence in 1980 when teams in the U.S.A. and the Soviet Union both reported evidence of nonzero neutrino rest masses. While these claims did not stand up, a new round of experiments once again indicates that $m_\nu$ (and hence $\Omega_\nu$) > 0.

The neutrino number density $n_\nu$ per species is 3/11 that of the CMB photons. Since the latter are in thermal equilibrium, their number density is $n_{\text{cmb}} = 2\zeta(3)(kT_{\text{cmb}}/\hbar c)^3/\pi^2$ [125] where $\zeta(3) = 1.202$. Multiplying by 3/11 and dividing through by the critical density (2.22), we obtain

$$\Omega_\nu = \frac{\sum m_\nu c^2}{(93.8 \text{ eV})h_0^2} , \tag{4.5}$$

where the sum is over three neutrino species. We follow convention and specify particle masses in units of $eV/c^2$, where $1 eV/c^2 = 1.602 \times 10^{-12} erg/c^2 = 1.783 \times 10^{-33}$ g. The calculations in this section are strictly valid only for $m_\nu c^2 \lesssim 1$ MeV. More massive neutrinos with $m_\nu c^2 \sim 1$ GeV were once considered as CDM candidates, but are no longer viable since experiments at the LEP collider rule out additional neutrino species with masses up to half that of the $Z_0$ ($m_{Z_0} c^2 = 91$ GeV).

Current laboratory upper bounds on neutrino rest masses are $m_{\nu_e} c^2 < 3$ eV, $m_{\nu_\mu} c^2 < 0.19$ MeV and $m_{\nu_\tau} c^2 < 18$ MeV, so it would appear feasible in principle for these particles to close the Universe. In fact $m_{\nu_\mu}$ and $m_{\nu_\tau}$ are limited far more stringently by (4.5) than by laboratory bounds. Perhaps the best-known theory along these lines is that of D.W. Sciama [126], who postulated a population of $\tau$-neutrinos with $m_{\nu_\tau} c^2 \approx 29$ eV. Eq. (4.5) shows that such neutrinos would account for much of the dark matter, contributing a *minimum* collective density of $\Omega_\nu \geqslant 0.38$ (assuming as usual that $h_0 \leqslant 0.9$). We will consider decaying neutrinos further in Chap. 7.

Strong upper limits can be set on $\Omega_\nu$ within the context of the gravitational instability picture. Neutrinos constitute *hot dark matter* (i.e. they are relativistic when they decouple from the primordial fireball) and are therefore able to stream freely out of density perturbations in the early Universe, erasing them before they have a chance to grow and suppressing the power spectrum $P(k)$ of density fluctuations on small scales $k$. Agreement between LSS theory and observation can be achieved in models with $\Omega_\nu$ as high as 0.2, but only in combination with values of the other cosmological parameters that are no longer considered realistic (e.g. $\Omega_{bar} + \Omega_{cdm} = 0.8$ and $h_0 = 0.5$) [127]. A recent 95% confidence-level upper limit on the neutrino density based on data from the 2dF galaxy survey [128] reads

$$\Omega_\nu < 0.24(\Omega_{bar} + \Omega_{cdm}) . \tag{4.6}$$

This is if no prior assumptions are made about the values of $\Omega_{m,0}$, $\Omega_{bar}$, $h_0$ and $n$. When combined with Eqs. (4.3) and (4.4) for $\Omega_{bar}$ and $\Omega_{cdm}$, this result implies that $\Omega_\nu < 0.15$. Thus if structure grows by gravitational instability as generally assumed, then neutrinos may still play a significant, but not dominant role in cosmological dynamics. Note that neutrinos lose energy after decoupling and become *non*relativistic on timescales $t_{nr} \approx 190,000$ yr $(m_\nu c^2/eV)^{-2}$ [129], so Eq. (4.6) is consistent with our neglect of relativistic particles in the Friedmann-Lemaître equation (Sec. 2.3).

*Lower* limits on $\Omega_\nu$ follow from a recent series of neutrino experiments employing particle accelerators (LSND [130]), cosmic rays in the upper atmosphere (Super-Kamiokande [131]), the flux of neutrinos from the Sun

(SAGE [132], Homestake [133], GALLEX [134], SNO [135]), nuclear reactors (KamLAND [136]) and directed neutrino beams (K2K [137]). The evidence in each case points to interconversions between species, known as neutrino oscillations, which can only take place if all species involved have nonzero rest masses. It now appears that oscillations occur between at least two neutrino mass eigenstates, whose masses squared differ by $\Delta_{21}^2 \equiv |m_2^2 - m_1^2| = 6.9_{-0.8}^{+1.5} \times 10^{-5}$ eV$^2/c^4$ and $\Delta_{31}^2 \equiv |m_3^2 - m_1^2| = 2.3_{-0.9}^{+0.7} \times 10^{-3}$ eV$^2/c^4$ [138]. Scenarios involving a fourth "sterile" neutrino were once thought to be needed but are disfavored by global fits to all the data. Oscillation experiments are sensitive to mass differences, and cannot fix the mass of any one neutrino flavor unless they are combined with another experiment such as neutrinoless double-beta decay [139]. Nevertheless, if neutrino masses are hierarchical, like those of other fermions, then one can take $m_3 \gg m_2 \gg m_1$ so that the above measurements impose a lower limit on total neutrino mass: $\Sigma m_\nu c^2 > 0.045$ eV. Putting this number into (4.5), we find that

$$\Omega_\nu > 0.0005 \,, \tag{4.7}$$

with $h_0 \leqslant 0.9$ as usual. If, instead, neutrino masses are nearly degenerate, then $\Omega_\nu$ could in principle be larger than this, but will in any case still lie below the upper bound (4.6) imposed by structure formation. The neutrino contribution to $\Omega_{tot,0}$ is thus anywhere from about one-tenth that of luminous matter (Sec. 4.2) to about one-quarter of that attributed to CDM (Sec. 4.3). We emphasize that, if $\Omega_{cdm}$ is small, then $\Omega_\nu$ must be small also. In theories (like that to be discussed in Chap. 7) where the density of neutrinos exceeds that of other forms of matter, one would need to modify the standard gravitational instability picture by encouraging the growth of structure in some other way, as for instance by "seeding" with cosmic strings.

## 4.5  Dark energy

There are at least four reasons to include a cosmological constant ($\Lambda$) in Einstein's field equations. The first is mathematical: $\Lambda$ plays a role in these equations similar to that of the additive constant in an indefinite integral [48]. The second is dimensional: $\Lambda$ provides a fundamental length scale for cosmology via the relation $R_\Lambda \equiv \Lambda^{-1/2}$ [46, 140]. The third is dynamical: $\Lambda$ determines the asymptotic expansion rate of the Universe according to Eq. (2.32). And the fourth is material: $\Lambda$ is related to the energy density of the vacuum via Eq. (2.21).

Table 4.1 Theoretical Estimates of $\Omega_{\Lambda,0}$

| Theory | Predicted value of $\rho_\Lambda$ | $\Omega_{\Lambda,0}$ |
|--------|-----------------------------------|----------------------|
| QCD | $(0.3 \text{ GeV})^4 \hbar^{-3} c^{-5} = 10^{16}$ g cm$^{-3}$ | $10^{44} h_0^{-2}$ |
| EW | $(200 \text{ GeV})^4 \hbar^{-3} c^{-5} = 10^{26}$ g cm$^{-3}$ | $10^{55} h_0^{-2}$ |
| GUTs | $(10^{19} \text{ GeV})^4 \hbar^{-3} c^{-5} = 10^{93}$ g cm$^{-3}$ | $10^{122} h_0^{-2}$ |

With so many reasons to take this term seriously, why was it ignored for so long? Einstein himself set $\Lambda = 0$ in 1931 "for reasons of logical economy," because he saw no hope of measuring this quantity experimentally at the time. He is often quoted as adding that its introduction in 1915 was the "biggest blunder" of his life. This comment (which was attributed to him by G. Gamow [141] but does not appear anywhere in his writings), is sometimes interpreted as a rejection of the very idea of a cosmological constant. It more likely represents Einstein's rueful recognition that, by invoking the $\Lambda$-term solely to obtain a static solution of the field equations, he had narrowly missed what would surely have been one of the greatest *triumphs* of his life: the prediction of cosmic expansion.

The relation between $\Lambda$ and the energy density of the vacuum has led to a quandary in more recent times: modern quantum field theories such as quantum chromodynamics (QCD), electroweak (EW) and grand unified theories (GUTs) imply unacceptably large values for $\rho_\Lambda$ and $\Omega_{\Lambda,0}$ (Table 4.1). This "cosmological-constant problem" has been reviewed by many people, but there is no consensus on how to solve it [142]. It is undoubtedly another reason why some cosmologists have preferred to set $\Lambda = 0$, rather than deal with a parameter whose physical origins are still unclear.

Setting $\Lambda$ to zero, however, is no longer an appropriate response because observations (reviewed below) now indicate that $\Omega_{\Lambda,0}$ is in fact of order *unity*. The cosmological constant problem has therefore become more baffling than before, in that an explanation of this parameter must apparently contain a cancellation mechanism which is almost—but not quite—exact, failing at precisely the 123rd decimal place.

One suggestion for understanding the possible nature of such a cancellation has been to treat the vacuum energy field literally as an Olbers-type summation of contributions from different places in the Universe [143]. It can then be handled with the same formalism that we have developed in Chaps. 2 and 3 for background radiation. This has the virtue of framing the problem in concrete terms, and raises some interesting possibilities, but does not in itself explain why the energy density inherent in such a field does not gravitate in the conventional way [144].

Another idea is that theoretical expectations for the value of $\Lambda$ might refer only to the latter's "bare" value, which could have been progressively "screened" over time. The cosmological constant then becomes a *variable* cosmological term [145]. In such a scenario the "low" value of $\Omega_{\Lambda,0}$ merely reflects the fact that the Universe is old. In general, however, this means modifying Einstein's field equations and/or introducing new forms of matter such as scalar fields. We look at this suggestion in more detail in Chap. 5.

A third possibility occurs in higher-dimensional gravity, where the cosmological constant can arise as an artefact of dimensional reduction (i.e. in extracting the appropriate four-dimensional limit from the theory). In such theories the "effective" $\Lambda_4$ may be small while its $N$-dimensional analog $\Lambda_N$ may be large [146]. We consider some aspects of higher-dimensional gravity in Chap. 9.

As a last recourse, some workers have argued that $\Lambda$ might be small "by definition," in the sense that a universe in which $\Lambda$ was large would be incapable of giving rise to intelligent observers like ourselves [147]. This is an application of the anthropic principle whose status, however, remains unclear.

## 4.6 Cosmological concordance

Let us pass to what is known about the value of $\Omega_{\Lambda,0}$ from cosmology. It is widely believed that the Universe originated in a big-bang singularity, rather than passing through a "big bounce" at the beginning of the current expansionary phase. Differentiating the Friedmann-Lemaître equation (2.31) with respect to time and setting both the expansion rate and its time derivative to zero leads to an upper limit (sometimes called the Einstein limit $\Omega_{\Lambda,E}$) on $\Omega_{\Lambda,0}$ as a function of $\Omega_{m,0}$. For $\Omega_{m,0} = 0.3$ the requirement that $\Omega_{\Lambda,0} < \Omega_{\Lambda,E}$ implies $\Omega_{\Lambda,0} < 1.71$, a limit that tightens to $\Omega_{\Lambda,0} < 1.16$ for $\Omega_{m,0} = 0.03$ [39].

A slightly stronger constraint can be formulated (for closed models) in terms of the antipodal redshift. The antipodes are the set of points located at $\chi = \pi$, where $\chi$ (radial coordinate distance) is related to $r$ by $d\chi = (1 - kr^2)^{-1/2} dr$. Using (2.7) and (2.12) this can be rewritten in the form $d\chi = -(c/H_0 R_0) dz/\tilde{H}(z)$ and integrated with the help of (2.31). Gravitational lensing of sources beyond the antipodes cannot give rise to normal (multiple) images [148], so the redshift $z_a$ of the antipodes must exceed that of the most distant normally-lensed object, currently an early star-forming galaxy at $z = 7$ [149]. Requiring that $z_a > 7$ leads to the upper bound $\Omega_{\Lambda,0} < 1.54$ if $\Omega_{m,0} = 0.3$. This tightens to $\Omega_{\Lambda,0} < 1.14$ for $\Lambda$BDM-type models with $\Omega_{m,0} = 0.03$.

The statistics of gravitational lenses lead to a different and stronger upper limit which applies regardless of geometry. The increase in path length for a given redshift in vacuum-dominated models (relative to, say, EdS) means that there are more sources to be lensed, and presumably more lensed objects to be seen. The observed frequency of lensed quasars, however, is rather modest, leading to a bound of $\Omega_{\Lambda,0} < 0.66$ for flat models [150]. This limit would be compromised if it should turn out that dust hides hide distant sources [151]. But radio lenses would be far less affected, and these give only slightly weaker constraints: $\Omega_{\Lambda,0} < 0.73$ (for flat models) or $\Omega_{\Lambda,0} \lesssim 0.4 + 1.5\Omega_{m,0}$ (for curved ones) [152]. Recent indications are that this method loses much of its sensitivity to $\Omega_{\Lambda,0}$ when assumptions about the lensing population are properly normalized to galaxies at high redshift [153]. A recent limit from radio lenses in the Cosmic Lens All-Sky Survey (CLASS) is $\Omega_{\Lambda,0} < 0.89$ for flat models [154].

Tentative *lower* limits have been set on $\Omega_{\Lambda,0}$ using faint galaxy number counts. This premise is similar to that behind lensing statistics: the enhanced comoving volume at large redshifts in vacuum-dominated models should lead to greater (projected) galaxy number densities at faint magnitudes. In practice, it has proven difficult to disentangle this effect from galaxy luminosity evolution. Early claims of a best fit at $\Omega_{\Lambda,0} \approx 0.9$ [155] have been disputed on the basis that the steep increase seen in numbers of blue galaxies is not matched in the K-band, where luminosity evolution should be less important [156]. Attempts to account more fully for evolution have subsequently led to a limit of $\Omega_{\Lambda,0} > 0.53$ [157]. More recently a reasonable fit has been demonstrated (for flat models) with a vacuum density parameter of $\Omega_{\Lambda,0} = 0.8$ [79].

Other evidence for a significant $\Omega_{\Lambda,0}$-term has come from numerical simulations of large-scale structure formation. Fig. 4.3 shows the evolution of massive structures between $z = 3$ and $z = 0$ in simulations by the VIRGO Consortium [158]. The $\Lambda$CDM model (top row) provides a qualitatively better match to the observed distribution of galaxies than EdS ("SCDM," bottom row). The improvement is especially marked at higher redshifts (left-hand panels). Power-spectrum analysis, however, reveals that the match is not particularly good in either case [158]. This could reflect *bias* (i.e. a systematic discrepancy between the distributions of mass and

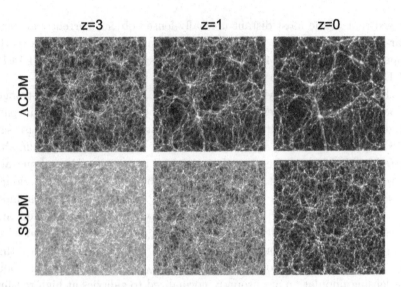

Fig. 4.3  Numerical simulations of structure formation. In the top row is the $\Lambda$CDM model with $\Omega_{m,0} = 0.3, \Omega_{\Lambda,0} = 0.7$ and $h_0 = 0.7$. The bottom row shows the EdS ("SCDM") model with $\Omega_{m,0} = 1, \Omega_{\Lambda,0} = 0$ and $h_0 = 0.5$. The panel size is comoving with the Hubble expansion, and time runs from left ($z = 3$) to right ($z = 0$). (Images courtesy of J. Colberg and the VIRGO Consortium.)

light). Different combinations of $\Omega_{m,0}$ and $\Omega_{\Lambda,0}$ might also provide better fits [47, 87, 159].

The first measurements to put both lower *and* upper bounds on $\Omega_{\Lambda,0}$ have come from type Ia supernovae (SNIa). These objects are very bright, with luminosities that are consistent (when calibrated against rise time), and they are not thought to evolve significantly with redshift. All of these properties make them ideal standard candles for use in the magnitude-redshift relation. In 1998 and 1999, two independent groups (HzT [160] and SCP [161]) reported a systematic dimming of SNIa at $z \approx 0.5$ by about 0.25 magnitudes relative to that expected in an EdS model, suggesting that space at these redshifts is "stretched" by dark energy. These programs have now expanded to encompass more than 200 supernovae, with results that can be summarized in the form of a 95% confidence-level relation between $\Omega_{m,0}$ and $\Omega_{\Lambda,0}$ [162]:

$$\Omega_{\Lambda,0} = 1.4\,\Omega_{m,0} + 0.35 \pm 0.28 \,. \tag{4.8}$$

Such a relationship is inconsistent with the EdS and OCDM models, which have large values of $\Omega_{m,0}$ with $\Omega_{\Lambda,0} = 0$. To extract quantitative limits on $\Omega_{\Lambda,0}$ alone, we recall that $\Omega_{m,0} \geqslant \Omega_{\text{bar}} \geqslant 0.02$ (Sec. 4.2) and $\Omega_{m,0} \leqslant 0.6$

(Sec. 4.3). Interpreting these constraints as conservatively as possible (i.e. $\Omega_{m,0} = 0.31 \pm 0.29$), we infer from (4.8) that

$$\Omega_{\Lambda,0} = 0.78 \pm 0.49 . \tag{4.9}$$

This is not a high-precision measurement, but it is enough to establish that $\Omega_{\Lambda,0} \geqslant 0.29$ and hence that *the dark energy is real*. Additional supernovae observations continue to reinforce this conclusion [163]. Not all cosmologists are yet convinced, however, and a healthy degree of caution is still in order regarding a quantity whose physical origin is so poorly understood. Alternative explanations can be constructed that fit the observations by means of "grey dust" [164] or luminosity evolution [165], though these must be increasingly fine-tuned to match recent SNIa data at higher redshifts [166]. Much also remains to be learned about the physics of supernova explosions. The shape of the magnitude-redshift relation suggests that observations may have to reach $z \sim 2$ routinely in order to be able to discriminate statistically between models (like $\Lambda$CDM and $\Lambda$BDM) with different ratios of $\Omega_{m,0}$ to $\Omega_{\Lambda,0}$.

Further support for the existence of dark energy has arisen from a completely independent source: the angular power spectrum of CMB fluctuations. These are produced by density waves in the primordial plasma, "the oldest music in the Universe" [167]. The first peak in the power spectrum picks out the angular size of the largest fluctuations in this plasma at the moment when the Universe became transparent to light. Because it is seen through the "lens" of a curved Universe, the location of this peak is sensitive to the latter's total density $\Omega_{\text{tot},0} = \Omega_{\Lambda,0} + \Omega_{m,0}$. Beginning in 2000, a series of increasingly precise measurements of $\Omega_{\text{tot},0}$ have been reported from experiments including BOOMERANG [168], MAXIMA [169], DASI [170] and most recently the WMAP satellite. The latter's results can be summarized as follows [171] (at the 95% confidence level, assuming $h_0 > 0.5$):

$$\Omega_{m,0} + \Omega_{\Lambda,0} = 1.03 \pm 0.05 . \tag{4.10}$$

*The Universe is therefore spatially flat, or very close to it.* To extract a value for $\Omega_{\Lambda,0}$ alone, we can do as in the SNIa case and substitute our matter density bounds ($\Omega_{m,0} = 0.31 \pm 0.29$) into (4.10) to obtain

$$\Omega_{\Lambda,0} = 0.72 \pm 0.29 . \tag{4.11}$$

This is consistent with (4.9), but has error bars which have been reduced by almost half, and are now due entirely to the uncertainty in $\Omega_{m,0}$. This measurement is impervious to most of the uncertainties of the earlier ones, because it leapfrogs "local" systems whose interpretation is complex (supernovae, galaxies, and quasars), going directly back to the radiation-dominated era when physics was simpler. Eq. (4.11) is sufficient to establish that $\Omega_{\Lambda,0} \geqslant 0.43$, and hence that *dark energy not only exists, but probably dominates the energy density of the Universe.*

The CMB power spectrum favors vacuum-dominated models, but is not yet resolved with sufficient precision to discriminate (on its own) between models which have exactly the critical density (like $\Lambda$CDM) and those which are close to the critical density (like $\Lambda$BDM). As it stands, the location of the first peak in these data actually hints at "marginally closed" models, although the implied departure from flatness is not statistically significant and could be explained in other ways [172].

Much attention is focused on the higher-order peaks of the spectrum, which contain valuable clues about the matter component. Odd-numbered peaks are produced by regions of the primordial plasma which have been maximally compressed by infalling material, and even ones correspond to maximally rarefied regions which have rebounded due to photon pressure. A high baryon-to-photon ratio enhances the compressions and retards the rarefractions, thus suppressing the size of the second peak relative to the first. The strength of this effect depends on the fraction of baryons (relative to the more weakly-bound neutrinos and CDM particles) in the overdense regions. The BOOMERANG and MAXIMA data show an unexpectedly weak second peak. While there are a number of ways to account for this in $\Lambda$CDM models (e.g., by "tilting" the primordial spectrum), the data are fit most naturally by a "no-CDM" $\Lambda$BDM model with $\Omega_{\rm cdm} = 0$, $\Omega_{m,0} = \Omega_{\rm bar}$ and $\Omega_{\Lambda,0} \approx 1$ [173]. Models of this kind have been discussed for some time in connection with analyses of the Lyman-$\alpha$ forest of quasar absorption lines [87, 174]. The WMAP data show a stronger second peak and are fit by both $\Lambda$CDM and $\Lambda$BDM-type models [175]. Data from the upcoming PLANCK satellite should settle this issue.

The best constraints on $\Omega_{\Lambda,0}$ come from taking *both* the supernovae and microwave background results at face value, and substituting one into the other. This provides a valuable cross-check on the matter density, because the SNIa and CMB constraints are very nearly orthogonal in the $\Omega_{m,0}$-$\Omega_{\Lambda,0}$ plane (Fig. 4.4). Thus, forgetting about our conservative bounds on $\Omega_{m,0}$ and merely substituting (4.10) into (4.8), we find

$$\Omega_{\Lambda,0} = 0.75 \pm 0.12 \ . \tag{4.12}$$

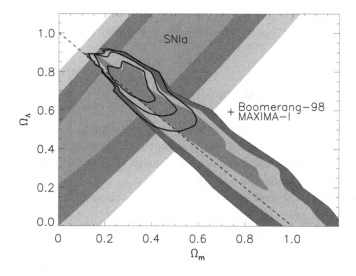

Fig. 4.4  Observational constraints on the values of $\Omega_{m,0}$ and $\Omega_{\Lambda,0}$ from both SNIa and CMB observations (BOOMERANG, MAXIMA). Shown are 68%, 95% and 99.7% confidence intervals inferred both separately and jointly from the data. (Reprinted from Ref. [177] by permission of A. Jaffe and P.L. Richards.)

Alternatively, extracting the matter density parameter, we obtain

$$\Omega_{m,0} = 0.28 \pm 0.12 . \tag{4.13}$$

These results further tighten the case for a universe dominated by dark energy. Eq. (4.12) also implies that $\Omega_{\Lambda,0} \leqslant 0.87$, which begins to put pressure on models of the $\Lambda$BDM type. Perhaps most importantly, Eq. (4.13) establishes that $\Omega_{m,0} \geqslant 0.16$, which is inconsistent with $\Lambda$BDM and *requires the existence of CDM*. Moreover, the fact that the range of values picked out by (4.13) agrees so well with that derived in Sec. 4.3 constitutes solid evidence for the $\Lambda$CDM model in particular, and for the gravitational instability picture of large-scale structure formation in general.

The depth of the change in thinking that has been triggered by these developments on the observational side can hardly be exaggerated. Only a few years ago, it was still routine to set $\Lambda = 0$ and cosmologists had two main choices: the "one true faith" (flat, with $\Omega_{m,0} \equiv 1$), or the "reformed" (open, with individual believers being free to choose their own values near $\Omega_{m,0} \approx 0.3$). All this has been irrevocably altered by the CMB experiments. If there is a guiding principle now, it is no longer $\Omega_{m,0} \approx 0.3$, and certainly not $\Omega_{\Lambda,0} = 0$; it is $\Omega_{tot,0} \approx 1$ from the power spectrum of the CMB.

Cosmologists have been obliged to accept a $\Lambda$-term, and it is not so much a question of whether or not it dominates the energy budget of the Universe, but by *how much*.

## 4.7 The coincidental Universe

The observational evidence reviewed in the foregoing sections has led us into the corner of parameter space occupied by vacuum-dominated models with close to (or exactly) the critical density. The resulting picture is self-consistent, and agrees with nearly all the data. Major questions, however, remain on the theoretical side. Prime among these is the problem of the cosmological constant, which (as described above) is particularly acute in models with nonzero values of $\Lambda$, because one can no longer hope that a simple symmetry of nature will eventually be found which requires $\Lambda = 0$.

A related concern has to do with the *evolution* of the matter and dark-energy density parameters $\Omega_m$ and $\Omega_\Lambda$ over time. Eqs. (2.22) and (2.30) can be combined to give

$$\Omega_m(t) \equiv \frac{\rho_m(t)}{\rho_{\text{crit}}(t)} = \frac{\Omega_{m,0}}{\tilde{R}^3(t)\,\tilde{H}^2(t)} \qquad \Omega_\Lambda(t) \equiv \frac{\rho_\Lambda(t)}{\rho_{\text{crit}}(t)} = \frac{\Omega_{\Lambda,0}}{\tilde{H}^2(t)} \,. \qquad (4.14)$$

Here $\tilde{H}[z(t)]$ is given by (2.31) as usual and $z(t) = 1/\tilde{R}(t) - 1$ from (2.11). Eqs. (4.14) can be solved exactly for spatially flat models using Eqs. (2.52) and (2.53) for $\tilde{R}(t)$ and $\tilde{H}(t)$. At early times, dark energy is insignificant relative to matter ($\Omega_\Lambda \sim 0$ and $\Omega_m \sim 1$), but the situation is reversed at late times when $\Omega_\Lambda \sim 1$ and $\Omega_m \sim 0$. Results for the $\Lambda$CDM model are illustrated in Fig. 4.5(a).

What is remarkable in this figure is the location of the present in relation to the values of $\Omega_m$ and $\Omega_\Lambda$. *We have apparently arrived on the scene at the precise moment when these two parameters are in the midst of switching places.* This has come to be known as the coincidence problem, and S.M. Carroll [142] has described such a universe as "preposterous." He writes: "This scenario staggers under the burden of its unnaturalness, but nevertheless crosses the finish line well ahead of any of its competitors by agreeing so well with the data." Cosmology may be moving toward a position like that of particle physics, where a standard model accounts for all observed phenomena to high precision, but appears to be founded on a series of finely-tuned parameters which leave one with the distinct impression that the underlying reality has not yet been grasped.

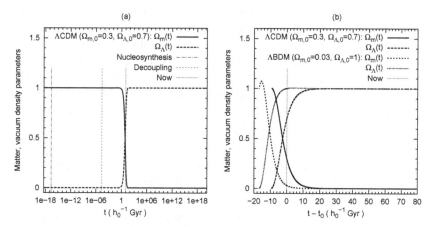

Fig. 4.5 The evolution of $\Omega_m(t)$ and $\Omega_\Lambda(t)$ in vacuum-dominated models. Panel (a) shows a single model ($\Lambda$CDM) over twenty powers of time in either direction. Plotted this way, we are seen to live at a very special time (marked "Now"). Standard nucleosynthesis ($t_{nuc} \sim 1$ s) and matter-radiation decoupling times ($t_{dec} \sim 10^{11}$ s) are included for comparison. Panel (b) shows both the $\Lambda$CDM and $\Lambda$BDM models on a linear rather than logarithmic scale, for the first $100h_0^{-1}$ Gyr after the big bang (i.e. the lifetime of the stars and galaxies).

Fig. 4.5(b) is a close-up view of Fig. 4.5(a), with one difference: it is plotted on a *linear* scale in time for the first $100h_0^{-1}$ Gyr after the big bang, rather than a logarithmic scale over $10^{\pm 20}h_0^{-1}$ Gyr. The rationale for this is simple: 100 Gyr is approximately the lifetime of the galaxies (as determined by their main-sequence stellar populations). One would not, after all, expect observers to appear on the scene long after all the galaxies had disappeared, or in the early stages of the expanding fireball. Seen from the perspective of Fig. 4.5(b), the coincidence, while still striking, is perhaps no longer so preposterous. The $\Lambda$CDM model still appears fine-tuned, in that "Now" follows rather quickly on the heels of the epoch of matter-vacuum equality. In the $\Lambda$BDM model, $\Omega_{m,0}$ and $\Omega_{\Lambda,0}$ are closer to the cosmological time-averages of $\Omega_m(t)$ and $\Omega_\Lambda(t)$ (namely zero and one respectively). In such a picture it might be easier to believe that we have not arrived on the scene at a special time, but merely a *late* one. Whether or not this is a helpful way to approach the coincidence problem is, to a certain extent, a matter of taste.

To summarize the contents of this chapter: what can be seen with our telescopes constitutes no more than one percent of the density of the Universe. The rest is dark. A small portion (no more than five percent) of this dark matter is made up of ordinary baryons. Many observational arguments hint at the existence of a second, more exotic species known as cold dark matter (though they do not quite establish its existence unless they are combined with ideas about the formation of large-scale structure). Experiments also imply the existence of a third dark-matter species, the massive neutrino, although its role appears to be more limited. Finally, all these components are dwarfed in importance by a newcomer, whose physical origin remains shrouded in obscurity: the dark energy of the vacuum.

In the remainder of this book, we explore the leading candidates for dark energy and matter in more detail: the cosmological vacuum, elementary particles such as axions, neutrinos and weakly-interacting massive particles (WIMPs), and black holes. Our approach is to constrain each proposal by calculating its contributions to the background radiation, and comparing these with observational data at wavelengths ranging from radio waves to gamma-rays. In the spirit of Olbers' paradox and what we have done so far, our question for each candidate will be: Just *how* dark is it?

Chapter 5

# The Radio and Microwave Backgrounds

## 5.1 The cosmological "constant"

The cosmological-constant problem is essentially the problem of reconciling the very high vacuum-energy densities expected on the basis of quantum field theory with the small (but nonzero) dark-energy density now inferred from cosmological observation (Sec. 4.5). Many authors have sought to bridge the gap by looking for a mechanism that would allow the energy density $\rho_v$ of the vacuum to *decay* with time. Since $\Lambda c^2 = 8\pi G \rho_v$ from (2.21), this means replacing Einstein's cosmological constant by a variable "cosmological term." With such a mechanism in hand, the problem would be reduced to explaining why the Universe is of intermediate age: old enough that $\Lambda$ has relaxed from primordial values like those suggested by quantum field theory to the values which we measure now, but young enough that $\Omega_\Lambda \equiv \rho_v/\rho_{\rm crit}$ has not yet reached its asymptotic value of unity.

Energy conservation requires that any decrease in the energy density of the vacuum be made up by a corresponding increase somewhere else. In some scenarios, dark energy goes into the kinetic energy of new forms of matter such as scalar fields, which have yet to be observed in nature. In others it is channelled instead into baryons, photons or neutrinos. Baryonic decays would produce equal amounts of matter and antimatter, whose subsequent annihilation would flood the Universe with gamma-rays. Radiative decays would similarly pump photons into intergalactic space, but are harder to constrain because they could in principle involve any part of the electromagnetic spectrum. As we will see, however, robust limits can be set on any such process under conservative assumptions.

But how can $\Lambda$, originally introduced by Einstein in 1917 as a constant of nature akin to $c$ and $G$, be allowed to vary? To answer this, we go back to the field equations of general relativity:

$$\mathcal{R}_{\mu\nu} - \frac{1}{2}\mathcal{R}\,g_{\mu\nu} - \Lambda\,g_{\mu\nu} = -(8\pi G/c^4)\,\mathcal{T}_{\mu\nu}\;. \tag{5.1}$$

The covariant derivative of these equations can be written in the following form with the help of the Bianchi identities, which read $\nabla^\nu(\mathcal{R}_{\mu\nu} - \frac{1}{2}\mathcal{R}\,g_{\mu\nu}) = 0$:

$$\partial_\mu\Lambda = \frac{8\pi G}{c^4}\,\nabla^\nu\mathcal{T}_{\mu\nu}\;. \tag{5.2}$$

Within Einstein's theory, it follows that $\Lambda = $ constant as long as matter and energy (as contained in $\mathcal{T}_{\mu\nu}$) are conserved.

In variable-$\Lambda$ theories, one must do one of three things: abandon matter-energy conservation, modify general relativity, or stretch the definition of what is conserved. The first of these routes was explored as early as 1933 by M. Bronstein [177], who sought to connect energy non-conservation with the cosmological arrow of time [178]. Few physicists today would be willing to sacrifice energy conservation outright. Some, however, would be willing to modify general relativity, or to consider new forms of matter and energy. Historically, these two approaches have sometimes been seen as distinct, with one being a change to the geometry of nature while the other is concerned with the material content of the Universe. The modern tendency, however, is to regard them as equivalent. This viewpoint is best personified by Einstein, who in 1936 compared the left-hand (geometrical) and right-hand (matter) sides of his field equations to "fine marble" and "low-grade wooden" wings of the same house [179]. In a more complete theory, he argued, matter fields of all kinds would be seen to be just as geometrical as the gravitational one.

## 5.2   The scalar field

Let us see how this works in one of the oldest and simplest variable-$\Lambda$ theories: a modification of general relativity in which the metric tensor $g_{\mu\nu}$ is supplemented by a scalar field $\varphi$ whose coupling to matter is determined by a parameter $\omega$. Ideas of this kind go back to P. Jordan in 1949 [180], M. Fierz in 1956 [181] and C. Brans and R.H. Dicke in 1961 [182]. Theorists then were not as casual about invoking hypothetical new scalar fields as they are now, and all these authors sought to associate $\varphi$ with a known quantity. Various lines of argument (notably Mach's principle) pointed to an identification with Newton's gravitational "constant" such that $G \sim 1/\varphi$. By 1968 it was appreciated that $\Lambda$ and $\omega$ too would depend on $\varphi$ in general [183]. The original Brans-Dicke theory (with $\Lambda = 0$) has subsequently been extended to *generalized scalar-tensor theories* in which $\Lambda = \Lambda(\varphi)$ [184], $\Lambda = \Lambda(\varphi)$, $\omega = \omega(\varphi)$ [185] and $\Lambda = \Lambda(\varphi, \psi)$, $\omega = \omega(\varphi)$ where $\psi \equiv \partial^\mu \varphi \, \partial_\mu \varphi$ [186]. In the last and most general of these cases, the field equations read

$$\mathcal{R}_{\mu\nu} - \frac{1}{2} \mathcal{R} \, g_{\mu\nu} + \frac{1}{\varphi} \left[ \nabla_\mu (\partial_\nu \varphi) - \Box \, \varphi \, g_{\mu\nu} \right] + \frac{\omega(\varphi)}{\varphi^2} \left( \partial_\mu \varphi \, \partial_\nu \varphi - \frac{1}{2} \psi \, g_{\mu\nu} \right)$$
$$- \Lambda(\varphi, \psi) \, g_{\mu\nu} + 2 \frac{\partial \Lambda(\varphi, \psi)}{\partial \psi} \partial_\mu \varphi \, \partial_\nu \varphi = -\frac{8\pi}{\varphi \, c^4} \, T_{\mu\nu} \, , \quad (5.3)$$

where $\Box \varphi \equiv \nabla^\mu (\partial_\mu \varphi)$ is the D'Alembertian. These reduce to Einstein's equations (5.1) when $\varphi = \text{const} = 1/G$.

If we now repeat the exercise on the previous page and take the covariant derivative of the field equations (5.3) with the Bianchi identities, we obtain a generalized version of the equation (5.2) faced by Bronstein:

$$\partial_\mu \varphi \left\{ \frac{\mathcal{R}}{2} + \frac{\omega(\varphi)}{2\varphi^2} \psi - \frac{\omega(\varphi)}{\varphi} \Box \, \varphi + \Lambda(\varphi, \psi) + \varphi \frac{\partial \Lambda(\varphi, \psi)}{\partial \varphi} - \frac{\psi}{2\varphi} \frac{d\omega(\varphi)}{d\varphi} \right.$$
$$\left. -2\varphi \Box \, \varphi \frac{\partial \Lambda(\varphi, \psi)}{\partial \psi} - 2\partial^\kappa \varphi \, \partial_\kappa \left[ \varphi \frac{\partial \Lambda(\varphi, \psi)}{\partial \psi} \right] \right\} = \frac{8\pi}{c^4} \nabla^\nu T_{\mu\nu} \, . \quad (5.4)$$

Now energy conservation ($\nabla^\nu T_{\mu\nu} = 0$) no longer requires $\Lambda = \text{const}$. In fact, it is generally *incompatible* with constant $\Lambda$, unless an extra condition is imposed on the terms inside the curly brackets in (5.4). Similar conclusions hold for other scalar-tensor theories in which $\varphi$ is no longer associated with $G$. Examples include models with non-minimal couplings between $\varphi$ and the curvature scalar $\mathcal{R}$ [187], conformal rescalings of the metric tensor by functions of $\varphi$ [188] and the inclusion of extra potentials $V(\varphi)$ for the scalar field [189–191]. (Theories of this last kind are now known as *quintessence* models [192]). In each of these scenarios, the cosmological "constant" becomes a dynamical variable.

In the modern approach to variable-$\Lambda$ cosmology, which goes back to Y.B. Zeldovich in 1968 [193], all extra terms of the kind just described— including $\Lambda$—are moved to the right-hand side of the field equations (5.3), leaving only the Einstein tensor $\mathcal{R}_{\mu\nu} - \frac{1}{2} \mathcal{R} \, g_{\mu\nu}$ on the "geometrical" left-hand side. The cosmological term, along with scalar (or other new) fields, are thus interpreted as *new kinds of matter*. Eqs. (5.3) become

$$\mathcal{R}_{\mu\nu} - \frac{1}{2} \mathcal{R} \, g_{\mu\nu} = -\frac{8\pi}{\varphi \, c^4} \, T_{\mu\nu}^{\text{eff}} + \Lambda(\varphi) \, g_{\mu\nu} \, . \quad (5.5)$$

Here $T_{\mu\nu}^{\text{eff}}$ is an effective energy-momentum tensor describing the sum of ordinary matter plus whatever scalar (or other) fields have been added to the theory. For generalized scalar-tensor theories as described above, this could be written as $T_{\mu\nu}^{\text{eff}} \equiv T_{\mu\nu} + T_{\mu\nu}^{\varphi}$ where $T_{\mu\nu}$ refers to ordinary matter and $T_{\mu\nu}^{\varphi}$ to the scalar field. For the case with $\Lambda = \Lambda(\varphi)$ and $\omega = \omega(\varphi)$, for instance, the latter would be defined by (5.3) as

$$T_{\mu\nu}^{\varphi} \equiv \frac{1}{\varphi}\left[\nabla_\mu(\partial_\nu\varphi) - \Box\,\varphi\,g_{\mu\nu}\right] + \frac{\omega(\varphi)}{\varphi^2}\left(\partial_\mu\varphi\,\partial_\nu\varphi - \frac{1}{2}\,\psi\,g_{\mu\nu}\right) . \qquad (5.6)$$

The covariant derivative of the field equations (5.5) now reads

$$0 = \nabla^\nu\left[\frac{8\pi}{\varphi\,c^4}\,T_{\mu\nu}^{\text{eff}} - \Lambda(\varphi)\,g_{\mu\nu}\right] . \qquad (5.7)$$

Eq. (5.7) carries the same physical content as (5.4), but is more general in form and can readily be extended to other theories. Physically, it says that energy *is* conserved in variable-$\Lambda$ cosmology—where "energy" is now understood to refer to the energy of ordinary matter along with that in any additional fields which may be present, *and along with that in the vacuum*, as represented by $\Lambda$. In general, the latter parameter can vary as it likes, so long as the conservation equation (5.7) is satisfied.

It was noticed by M. Endō and T. Fukui in 1977 [184] that theories of this kind address the cosmological-constant problem, in the sense that $\Lambda$ can drop from large primordial values to ones like those seen today. These authors found solutions for $\varphi(t)$ such that $\Lambda \propto t^{-2}$ when $\Lambda = \Lambda(\varphi)$ and $\omega = $ constant. Similarly, S.M. Barr [189] found models in which $\Lambda \propto t^{-\ell}$ at late times, while P.J.E. Peebles and B. Ratra [190] discussed a theory in which $\Lambda \propto R^{-m}$ at early ones (here $\ell$ and $m$ are powers). There is now a rich literature on $\Lambda$-decay laws of this kind and their implications for cosmology [145].

In assessing such models, however, two caveats must be kept in mind. The first is theoretical. Insofar as the mechanisms discussed so far are entirely classical, they do not address the underlying problem. For this, one would also need to explain why net contributions to $\Lambda$ from the *quantum vacuum* do not remain at the primordial level, or how they are suppressed with time. A.M. Polyakov [194] and S.M. Adler [195] in 1982 were the first to speculate explicitly that such a suppression might come about if the "bare" cosmological term implied by quantum field theory were progressively screened by an "induced" counterterm of opposite sign, driving the effective value of $\Lambda(t)$ toward zero at late times. Many theoretical adjustment mechanisms have now been identified as potential sources of such a screening effect, beginning with a 1983 suggestion by A.D. Dolgov [196] based on non-minimally coupled scalar fields. Subsequent proposals have involved scalar fields [197–199], fields of higher spin [200–202], quantum effects during inflation [203–205] and other phenomena [206–208]. In most of these cases, no analytic expression is found for $\Lambda$ in terms of time or other cosmological parameters; the intent is merely to demonstrate that decay (and preferably near-cancellation) of the cosmological term is possible in principle. None of these mechanisms has been widely accepted. In fact, there is a general argument due to S. Weinberg to the effect that a successful mechanism based on scalar fields would necessarily be so finely-tuned as to be just as mysterious as the original problem [209]. Similar concerns have been raised in the case of vector and tensor-based proposals [210]. Nevertheless, the idea of an adjustment mechanism remains feasible in principle, and continues to attract considerable attention as a potential solution to the cosmological-constant problem.

The second caution is empirical. Observational data place increasingly strong restrictions on the way in which $\Lambda$ can vary with time. Among the most important are early-time bounds on the vacuum-energy density $\rho_\Lambda c^2 = \Lambda c^4 / 8\pi G$. The success of standard big-bang nucleosynthesis theory implies that $\rho_\Lambda$ was smaller than $\rho_r$ and $\rho_m$ during the radiation-dominated era, and large-scale structure formation could not have proceeded in the conventional way unless $\rho_\Lambda < \rho_m$ during the early matter-dominated era. Since $\rho_r \propto R^{-4}$ and $\rho_m \propto R^{-3}$ from (2.30), these requirements mean in practice that the vacuum-energy density must climb *less steeply than* $R^{-3}$ in the past direction, if it is comparable to that of matter or radiation at present [211, 212]. The variable-$\Lambda$ term must also satisfy late-time bounds like those which have been placed on the cosmological constant by purely astrophysical means (Sec. 4.5). Tests of this kind have been carried out using data on the age of the Universe [213, 214], structure formation [215–217], galaxy number counts [218], the CMB power spectrum [219, 220], gravitational lensing statistics [220–222] and Type Ia supernovae [220, 223]. In some cases, variable-$\Lambda$ models may be compatible with a nonsingular "big bounce" rather than a big bang [224], a possibility which can be ruled out on general grounds if $\Lambda = $ constant.

A third group of limits comes from asking what the vacuum decays *into*. In quintessence theories, dark energy is transferred to the kinetic energy of a scalar field as it "rolls" down a gradient toward the minimum of its potential. This may have observable consequences if the scalar field is coupled strongly to ordinary matter, but is hard to constrain in general. A simpler situation is that in which the vacuum decays into known particles such as baryons, photons or neutrinos. The baryonic channel would produce excessive levels of gamma-ray background radiation due to matter-antimatter annihilation, unless the energy density of the vacuum component is less than $3 \times 10^{-5}$ times that of matter [211]. This limit can be weakened if the decay process violates baryon number, or if it takes place in such a way that matter and antimatter are segregated on large scales; but such conditions are hard to arrange in a natural way. The radiative channel is more promising, but must pass criteria based on thermodynamic stability [225] and adiabaticity [226]. The *shape* of the spectrum of decay photons must not differ too much from that of pre-existing background radiation, or distortions will arise. K. Freese *et al.* have argued on this basis that the energy density of a vacuum decaying primarily into low-energy photons could not exceed $4 \times 10^{-4}$ times that of radiation [211].

Vacuum-decay photons may, however, blend into the spectrum of background radiation without distorting it. Fig. 1.15 shows that the best place to "hide" such a process would be the microwave region, where the energy density of background radiation is highest. Could all or part of the CMB be due to dark-energy decay? We know from the COBE satellite that its spectrum is very nearly that of a perfect blackbody [227]. Freese *et al.* pointed out that vacuum-decay photons would be thermalized by brehmsstrahlung and double-Compton scattering in the early Universe, and might continue to assume a blackbody spectrum at later times if pre-existing CMB photons played a role in "inducing" the vacuum to decay [211]. Subsequent work has shown that this would require a special combination of thermodynamical parameters [228]. This possibility is important in practice, however, because it leads to the most conservative limits on the theory. Even if the radiation produced by decaying dark energy does not distort the background, it will contribute to the latter's *absolute intensity*. We can calculate the size of these contributions to the background radiation using the methods laid out in Chaps. 2 and 3.

## 5.3   Decaying dark energy

The first step in this problem is to solve the field equations and conservation equations for the energy density of the decaying vacuum. We will do this in the context of a general phenomenological model. This means that we retain the field equations (5.5) and the conservation law (5.7), without specifying the form of the effective energy-momentum tensor in terms of scalar (or other) fields. These equations may be written

$$\mathcal{R}_{\mu\nu} - \frac{1}{2}\mathcal{R}\,g_{\mu\nu} = -\frac{8\pi G}{c^4}\left(\mathcal{T}_{\mu\nu}^{\text{eff}} - \rho_\Lambda c^2\,g_{\mu\nu}\right) \qquad (5.8)$$

$$0 = \nabla^\nu\left(\mathcal{T}_{\mu\nu}^{\text{eff}} - \rho_\Lambda c^2\,g_{\mu\nu}\right) . \qquad (5.9)$$

Here $\rho_\Lambda c^2 \equiv \Lambda c^4/8\pi G$ from (2.21) and we have replaced $1/\varphi$ with $G$ (assumed to be constant in what follows). Eqs. (5.8) and (5.9) have the same form as their counterparts (5.1) and (2.27) in standard cosmology, the key difference being that the cosmological term has migrated to the right-hand side and is *no longer necessarily constant*. Its evolution is now governed by the conservation equations (5.9), which require only that any change in $\rho_\Lambda c^2 g_{\mu\nu}$ be balanced by an equal and opposite change in the energy-momentum tensor $\mathcal{T}_{\mu\nu}^{\text{eff}}$.

While the latter is model-dependent in general, it is reasonable to assume in the context of isotropic and homogeneous cosmology that its form is that of a perfect fluid, as given by (2.24):

$$\mathcal{T}_{\mu\nu}^{\text{eff}} = (\rho_{\text{eff}} + p_{\text{eff}}/c^2)U_\mu U_\nu + p_{\text{eff}}\,g_{\mu\nu} . \qquad (5.10)$$

Comparison of Eqs. (5.9) and (5.10) shows that the conserved quantity in (5.9) must then *also* have the form of a perfect-fluid energy-momentum tensor, with density and pressure given by

$$\rho = \rho_{\text{eff}} + \rho_\Lambda \qquad p = p_{\text{eff}} - \rho_\Lambda c^2 . \qquad (5.11)$$

The conservation law (5.9) may then be simplified by analogy with Eq. (2.27):

$$\frac{1}{R^3}\frac{d}{dt}\left[R^3\left(\rho_{\text{eff}}c^2 + p_{\text{eff}}\right)\right] = \frac{d}{dt}\left(p_{\text{eff}} - \rho_\Lambda c^2\right) . \qquad (5.12)$$

This reduces to the standard result (2.28) for the case of a constant cosmological term, $\rho_\Lambda = $ const. Throughout this chapter, we allow the cosmological term to contain both a constant part *and* a time-varying part:

$$\rho_\Lambda = \rho_c + \rho_v(t) \qquad \rho_c = \text{ const} . \qquad (5.13)$$

We assume in addition that the perfect fluid described by $\mathcal{T}_{\mu\nu}^{\text{eff}}$ consists of a mixture of dust-like matter ($p_m = 0$) and radiation ($p_r = \frac{1}{3}\rho_r c^2$):

$$\rho_{\text{eff}} = \rho_m + \rho_r \qquad p_{\text{eff}} = \frac{1}{3}\rho_r c^2 . \qquad (5.14)$$

The conservation equation (5.12) then reduces to

$$\frac{1}{R^4}\frac{d}{dt}\left(R^4\rho_r\right) + \frac{1}{R^3}\frac{d}{dt}\left(R^3\rho_m\right) + \frac{d\rho_v}{dt} = 0 . \qquad (5.15)$$

From this equation it is clear that one (or both) of the radiation and matter densities can no longer obey the usual relations $\rho_r \propto R^{-4}$ and $\rho_m \propto R^{-3}$ in a theory with $\Lambda \neq$ const. Any change in $\Lambda$ (or $\rho_\Lambda$) must be accompanied by a change in radiation and/or matter densities.

To go further, some simplifying assumptions must be made. Let us take to begin with:

$$\frac{d}{dt}\left(R^3 \rho_m\right) = 0 \,. \tag{5.16}$$

This is just conservation of particle number, as may be seen by replacing "galaxies" with "particles" in Eq. (2.4). Such an assumption is well justified during the matter-dominated era by the stringent constraints on matter creation discussed in Sec. 5.1. It is equally valid during the radiation-dominated era, when the matter density is small, so that the $\rho_m$ term is of secondary importance compared to the other terms in (5.15).

In light of Eqs. (5.15) and (5.16), the vacuum can exchange energy only with radiation. As a model for this process, we follow M.D. Pollock in 1980 [229] and assume that it takes place in such a way that the energy density of the decaying vacuum component remains proportional to that of radiation, $\rho_v \propto \rho_r$. We adopt the notation of Freese et al. (1987) and write the proportionality factor as $x/(1-x)$, with $x$ the coupling parameter of the theory [211]. If this is allowed to take (possibly different) constant values during the radiation and matter-dominated eras, then

$$x \equiv \frac{\rho_v}{\rho_r + \rho_v} = \begin{cases} x_r & (t < t_{\text{eq}}) \\ x_m & (t \geqslant t_{\text{eq}}) \,. \end{cases} \tag{5.17}$$

Here $t_{\text{eq}}$ refers to the epoch of matter-radiation equality when $\rho_r = \rho_m$. Standard cosmology is recovered in the limit $x \to 0$. The most natural situation is that in which the value of $x$ stays constant, so that $x_r = x_m$. However, since observational constraints on $x$ are in general different for the radiation and matter-dominated eras, the most conservative limits on the theory are obtained by letting $x_r$ and $x_m$ take different values. Physically, this would correspond to a phase transition or sudden change in the expansion rate $\dot{R}/R$ of the Universe at $t = t_{\text{eq}}$.

With Eqs. (5.16) and (5.17), the conservation equation (5.15) reads

$$\frac{\dot{\rho}_v}{\rho_v} + 4(1-x)\frac{\dot{R}}{R} = 0 \,, \tag{5.18}$$

where overdots denote derivatives with respect to time. Integration gives

$$\rho_v(R) = \alpha_v R^{-4(1-x)} \,, \tag{5.19}$$

where $\alpha_v$ is a constant. The cosmological term $\Lambda$ is thus an inverse power-law function of the scale factor $R$, a scenario that has been widely studied [145]. Eq. (5.19) shows that the conserved quantity in this theory has a form intermediate between that of ordinary radiation entropy $(R^4 \rho_r)$ and particle number $(R^3 \rho_m)$ when $0 < x < \frac{1}{4}$.

The fact that $\rho_r \propto \rho_v \propto R^{-4(1-x)}$ places an immediate upper limit of $\frac{1}{4}$ on $x$ (in both eras), since higher values would erase the dynamical distinction between radiation and matter. With $x \leqslant \frac{1}{4}$ it then follows from (5.17) that $\rho_v \leqslant \frac{1}{3}\rho_r$. Freese *et al.* [211] obtained a stronger bound by showing that $x \leqslant 0.07$ if the baryon-to-photon ratio $\eta$ is to be consistent with both big-bang nucleosynthesis and observations of the CMB. (This argument assumes that $x = x_r = x_m$.) As a guideline in what follows, then, we will allow $x_r$ and $x_m$ to take values between zero and 0.07, and consider in addition the theoretical possibility that $x_m$ could increase to 0.25 in the matter-dominated era.

## 5.4 Energy density

With $\rho_m(R)$ specified by (5.16), $\rho_r$ related to $\rho_v$ by (5.17) and $\rho_v(R)$ given by (5.19), we can solve for all three components as functions of time if the scale factor $R(t)$ is known. This comes as usual from the field equations (5.8). Since these are the same as Eqs. (5.1) for standard cosmology, they lead to the same result, Eq. (2.20):

$$\left(\frac{\dot{R}}{R}\right)^2 = \frac{8\pi G}{3}\left(\rho_m + \rho_r + \rho_v + \rho_c\right). \tag{5.20}$$

Here we have used Eqs. (5.13) to replace $\rho_\Lambda$ with $\rho_v + \rho_c$ and (5.14) to replace $\rho_{\text{eff}}$ with $\rho_m + \rho_r$. We have also set $k = 0$ since observations indicate that these components together make up very nearly the critical density (Chap. 4).

Eq. (5.20) can be solved analytically in the three cases are of greatest physical interest: (1) the *radiation-dominated regime*, for which $t < t_{\text{eq}}$ and $\rho_r + \rho_v \gg \rho_m + \rho_c$; (2) the *matter-dominated regime*, which has $t \geqslant t_{\text{eq}}$ and $\rho_r + \rho_v \ll \rho_m$ (if $\rho_c = 0$); and (3) the *vacuum-dominated regime*, for which $t \geqslant t_{\text{eq}}$ and $\rho_r + \rho_v \ll \rho_m + \rho_c$. We draw a distinction between regimes 2 and 3 in order to model both matter-dominated universes similar to EdS and vacuum-dominated cosmologies like $\Lambda$CDM or $\Lambda$BDM (Table 3.1). We use the terms "EdS," "$\Lambda$CDM" and "$\Lambda$BDM" in this chapter to refer to flat models in which $\Omega_{m,0} = 1$, 0.3 and 0.03 respectively. These models differ slightly from their namesakes elsewhere in the book, since radiation and decaying-vacuum components are also included; but the differences are small except at the earliest times. In all cases, the present dark-energy density (if any) comes almost entirely from its constant-density component.

Eqs. (5.16), (5.17), (5.19) and (5.20) can be solved analytically for $R, \rho_m, \rho_r$ and $\rho_v$ in terms of $R_0$, $\rho_{m,0}$, $\rho_{r,0}$, $x_r$ and $x_m$ [39]. The normalized scale factor is found to read

$$\tilde{R}(t) = \begin{cases} \left(\dfrac{t}{t_0}\right)^{1/2(1-x_r)} & (t < t_{\text{eq}}) \\ \left[\dfrac{\mathcal{S}_m(t)}{\mathcal{S}_m(t_0)}\right]^{2/3} & (t \geqslant t_{\text{eq}}) . \end{cases} \tag{5.21}$$

The dark-energy density is given by

$$\rho_v(t) = \begin{cases} \dfrac{\alpha x_r}{(1-x_r)^2}\, t^{-2} & (t < t_{\text{eq}}) \\ \left(\dfrac{x_m}{1-x_m}\right)\rho_r(t) & (t \geqslant t_{\text{eq}}) , \end{cases} \tag{5.22}$$

where $\alpha = 3/(32\pi G) = 4.47 \times 10^5$ g cm$^{-2}$ s$^2$. The densities of radiation and matter are

$$\rho_r(t) = \begin{cases} \left(\dfrac{1-x_r}{x_r}\right)\rho_v(t) & (t < t_{\text{eq}}) \\ \rho_{r,0}\left[\dfrac{\mathcal{S}_m(t)}{\mathcal{S}_m(t_0)}\right]^{-8(1-x_m)/3} & (t \geqslant t_{\text{eq}}) \end{cases} \tag{5.23}$$

$$\frac{\rho_m(t)}{\rho_{m,0}} = \begin{cases} \left[\dfrac{\mathcal{S}_m(t_{\text{eq}})}{\mathcal{S}_m(t_0)}\right]^{-2}\left(\dfrac{t}{t_{\text{eq}}}\right)^{-3/2(1-x_r)} & (t < t_{\text{eq}}) \\ \left[\dfrac{\mathcal{S}_m(t)}{\mathcal{S}_m(t_0)}\right]^{-2} & (t \geqslant t_{\text{eq}}) . \end{cases} \tag{5.24}$$

Here we have applied $\rho_{m,0} = \Omega_{m,0}\,\rho_{\text{crit},0}$ and $\rho_{r,0} = \Omega_{r,0}\,\rho_{\text{crit},0}$ as boundary conditions. The function $\mathcal{S}_m(t)$ is defined as

$$\mathcal{S}_m(t) \equiv \begin{cases} t & (\Omega_{m,0} = 1) \\ \sinh(t/\tau_0) & (0 < \Omega_{m,0} < 1) , \end{cases} \tag{5.25}$$

where $\tau_0 \equiv 2/(3H_0\sqrt{1-\Omega_{m,0}})$. The age of of the Universe is

$$t_0 = \begin{cases} 2/(3H_0) & (\Omega_{m,0} = 1) \\ \tau_0 \sinh^{-1}\chi_0 & (0 < \Omega_{m,0} < 1) , \end{cases} \tag{5.26}$$

where $\chi_0 \equiv \sqrt{(1-\Omega_{m,0})/\Omega_{m,0}}$ and we have used Eq. (2.54). Corrections from the radiation-dominated era can be ignored since $t_0 \gg t_{\text{eq}}$.

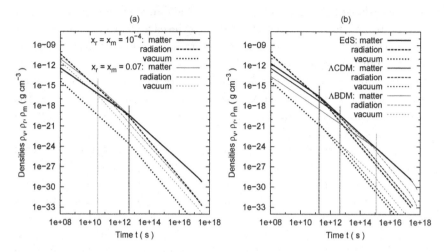

Fig. 5.1    The densities of decaying dark energy ($\rho_v$), radiation ($\rho_r$) and matter ($\rho_m$) as functions of time. Panel (a) shows the effects of changing the values of $x_r$ and $x_m$, assuming a model with $\Omega_{m,0} = 0.3$ (similar to $\Lambda$CDM). Panel (b) shows the effects of changing the cosmological model, assuming $x_r = x_m = 10^{-4}$. The vertical lines indicate the epochs when the densities of matter and radition are equal ($t_{eq}$). All curves assume $h_0 = 0.75$.

The parameter $t_{eq}$ is obtained as in standard cosmology by setting $\rho_r(t_{eq}) = \rho_m(t_{eq})$ in Eqs. (5.23). This leads to

$$
t_{eq} = \begin{cases} t_0\, \Omega_{r,0}^{3/2(1-4x_m)} & (\Omega_{m,0} = 1) \\ \tau_0 \sinh^{-1}\left[\chi_0\left(\dfrac{\Omega_{r,0}}{\Omega_{m,0}}\right)^{3/2(1-4x_m)}\right] & (0 < \Omega_{m,0} < 1)\,. \end{cases} \tag{5.27}
$$

This quantity is important because it is at about $t_{eq}$ that the Universe became transparent to radiation. Decay photons created before $t_{eq}$ would have been thermalized by the primordial plasma and eventually re-emitted as part of the CMB. It is the decay photons emitted *after* this time which can contribute to the extragalactic background radiation, and whose contributions we wish to calculate. The quantity $t_{eq}$ is thus analogous to the galaxy formation time $t_f$ in previous sections.

The densities $\rho_m(t)$, $\rho_r(t)$ and $\rho_v(t)$ are plotted as functions of time in Fig. 5.1. The left-hand panel (a) shows the effects of varying the parameters $x_r$ and $x_m$ within a given cosmological model (here, $\Lambda$CDM). Raising the value of $x_m$ leads to a proportionate increase in $\rho_v$ and a modest drop in $\rho_r$. It also flattens the slope of both components. The change in slope (relative to that of the matter component) pushes the epoch of equality back toward the big bang (vertical lines). Such an effect could in principle allow more time for structure to form during the early matter-dominated era [211], although the "compression" of the radiation-dominated era rapidly becomes unrealistic for values of $x_m$ close to $\frac{1}{4}$. Thus Fig. 5.1(a) shows that the value of $t_{eq}$ is reduced by a factor of over 100 in going from a model with $x_m = 10^{-4}$ to one with $x_m = 0.07$. In the limit $x_m \to \frac{1}{4}$, the radiation-dominated era disappears altogether, as remarked above and as shown explicitly by Eqs. (5.27).

Fig. 5.1(b) shows the effects of changes in cosmological model for fixed values of $x_r$ and $x_m$ (here both set to $10^{-4}$). Moving from the matter-filled EdS model toward vacuum-dominated ones such as $\Lambda$CDM and $\Lambda$BDM does three things. The first is to increase the age ($t_0$) of the Universe. This increases the density of radiation at any given time, since the latter is fixed at present and climbs at the same rate in the past direction. Based on our experience with the EBL, we might expect that this would lead to significantly higher levels of background radiation when integrated over time. However, there is a second effect at play here which acts in the opposite direction: smaller values of $\Omega_{m,0}$ boost the value of $t_{eq}$ as well as $t_0$, delaying the onset of the matter-dominated era (vertical lines). These two changes largely cancel each other out as far as dark-energy contributions to the background are concerned. The third effect of vacuum energy is an acceleration in the expansion rate at late times (Fig. 4.2), leading to a slight "droop" in the densities of all three components at the right-hand edge of Fig. 5.1(b) for the $\Lambda$CDM and $\Lambda$BDM models.

## 5.5   Source luminosity

In order to make use of the formalism we have developed in Chaps. 2 and 3, we need to define discrete "sources" of radiation from dark-energy decay, analogous to the galaxies of previous sections. For this purpose we carve up the Universe into hypothetical regions of arbitrary comoving volume $V_0$. The comoving number density of these source regions is just

$$n(t) = n_0 = V_0^{-1} = \text{const} . \tag{5.28}$$

The next step is to identify the "source luminosity." There are at least two ways to approach this question [230]. One could simply regard the source region as a ball of physical volume $V(t) = \tilde{R}^3(t)V_0$ filled with fluctuating dark energy. As the density of this energy drops by $-d\rho_v$ during time $dt$, the ball loses energy at a rate $-d\rho_v/dt$. If some fraction $\beta$ of this energy flux goes into photons, then the luminosity of the ball is

$$L_v(t) = -\beta\, c^2 \dot{\rho}_v(t)\, V(t) \ . \tag{5.29}$$

Such a definition is implicitly assumed in investigations of the stability of the vacuum-decay process based on the requirement that fluctuations in $\rho_v$ not grow larger than its mean value of with time [225]. For convenience we will refer to (5.29) as the *thermodynamical definition* of vacuum luminosity ($L_{th}$).

A second approach is to treat this as a problem involving spherical symmetry within general relativity. The assumption of spherical symmetry allows the total mass-energy ($Mc^2$) of a localized region of perfect fluid to be identified unambiguously. Luminosity can then be related to the *time rate of change* of this mass-energy. Assuming once again that the two are related by a factor $\beta$, we have

$$L_v(t) = \beta\, \dot{M}(t)\, c^2 \ . \tag{5.30}$$

Einstein's field equations then lead to the following expression [231] for the rate of change of mass-energy in terms of the pressure $p_v$ at the region's surface:

$$\dot{M}(t)\, c^2 = -4\pi\, p_v(t)\, [r(t)]^2 \dot{r}(t) \ , \tag{5.31}$$

where $r(t) = \tilde{R}(t)r_0$ is the region's physical radius. Taking $V = \frac{4}{3}\pi r^3$, applying the vacuum equation of state $p_v = -\rho_v c^2$ and substituting (5.31) into (5.30), we find that the latter can be written in the form

$$L_v(t) = \beta\, c^2 \rho_v(t)\, \dot{V}(t) \ . \tag{5.32}$$

This is just as appealing dimensionally as Eq. (5.29), and shifts the emphasis physically from fluctuations in the material content of the source region toward changes in its geometry. We will refer to (5.32) for convenience as the *relativistic definition* of vacuum luminosity ($L_{rel}$).

It is not obvious which of the two definitions (5.29) and (5.32) more correctly describes the luminosity of decaying dark energy; this is a conceptual issue. The ratio of the two reads

$$\frac{L_{th}}{L_{rel}} = -\frac{\dot{\rho}_v V}{\rho_v \dot{V}} = -\frac{1}{3}\frac{\dot{\rho}_v}{\rho_v}\frac{R}{\dot{R}} \ . \tag{5.33}$$

Differentiating Eqs. (5.21) and (5.22) with respect to time, we find

$$\frac{\dot{R}}{R} = \begin{cases} \dfrac{2}{3t} & (\Omega_{m,0} = 1) \\[2mm] \dfrac{2}{3\tau_0} \coth\left(\dfrac{t}{\tau_0}\right) & (0 < \Omega_{m,0} < 1) \end{cases} \tag{5.34}$$

$$\frac{\dot{\rho}_v}{\rho_v} = \begin{cases} -\dfrac{8}{3t}(1 - x_m) & (\Omega_{m,0} = 1) \\[2mm] -\dfrac{8}{3\tau_0}(1 - x_m)\coth\left(\dfrac{t}{\tau_0}\right) & (0 < \Omega_{m,0} < 1) \,. \end{cases} \tag{5.35}$$

The ratio $L_{\rm th}/L_{\rm rel} = 4(1 - x_m)/3$, ranging between values of $\frac{4}{3}$ (in the limit $x_m \to 0$ where standard cosmology is recovered) and 1 (in the opposite limit where $x_m$ takes its maximum theoretical value of $\frac{1}{4}$). There is thus little difference between the two scenarios in practice, at least where this model of decaying dark energy is concerned. We will proceed using the relativistic definition (5.32), which gives lower intensities and hence more conservative limits on the theory. At the end of the section we can check the corresponding intensity for the thermodynamical case (5.29) by simply multiplying through by $\frac{4}{3}(1 - x_m)$.

We now turn to the question of the branching ratio $\beta$, or fraction of decaying dark energy which goes into photons as opposed to other forms of radiation such as massless neutrinos. This is model-dependent in general. If the vacuum-decay radiation reaches equilibrium with that already present, however, then we may reasonably set this equal to the ratio of photon-to-total radiation energy densities in the CMB:

$$\beta = \Omega_\gamma/\Omega_{r,0} \,. \tag{5.36}$$

The density parameter $\Omega_\gamma$ of CMB photons is given in terms of their blackbody temperature $T$ by Stefan's law. With $T_{\rm cmb} = 2.728$ K [227] we get

$$\Omega_\gamma = \frac{4\sigma_{\rm SB}T^4}{c^3\rho_{\rm crit,0}} = 2.48 \times 10^{-5}h_0^{-2} \,. \tag{5.37}$$

The total radiation density $\Omega_{r,0} = \Omega_\gamma + \Omega_\nu$ is harder to determine, since there is little prospect of detecting the neutrino component directly. What is done in standard cosmology is to calculate the size of neutrino contributions to $\Omega_{r,0}$ under the assumption of entropy conservation. With three light neutrino species, this leads to

$$\Omega_{r,0} = \Omega_\gamma \left[1 + 3 \times \frac{7}{8}\left(\frac{T_\nu}{T}\right)^4\right] \,, \tag{5.38}$$

where $T_\nu$ is the blackbody temperature of the relic neutrinos [232]. During the early stages of the radiation-dominated era, neutrinos were in thermal equilibrium with photons so that $T_\nu = T$. They dropped out of equilibrium, however, when the temperature of the expanding fireball dropped below about $kT \sim 1$ MeV (the energy scale of weak interactions). Shortly thereafter, when the temperature dropped to $kT \sim m_e c^2 = 0.5$ MeV, electrons and positrons began to annihilate, transferring their entropy to the remaining photons in the plasma. This raised the photon temperature by a factor of $(1 + 2 \times \frac{7}{8} = \frac{11}{4})^{1/3}$ relative to that of the neutrinos. In standard cosmology, the ratio of $T_\nu/T$ has remained at $(4/11)^{1/3}$ down to the present day, so that (5.38) gives

$$\Omega_{r,0} = 1.68 \, \Omega_\gamma = 4.17 \times 10^{-5} h_0^{-2} \, . \tag{5.39}$$

Using (5.36) for $\beta$, this would imply:

$$\beta = 1/1.68 = 0.595 \, . \tag{5.40}$$

We will take these as our "standard values" of $\Omega_{r,0}$ and $\beta$ in what follows. They are conservative ones, in the sense that most alternative lines of argument would imply higher values of $\beta$. It has, for instance, been argued that vacuum decay with $x_r =$ const. would be easier to reconcile with processes such as electron-positron annihilation if the vacuum coupled to photons but not neutrinos [212]. This would complicate the theory, breaking the radiation density $\rho_r$ in (5.15) into a photon part $\rho_\gamma$ and a neutrino part $\rho_\nu$ with different dependencies on $R$. Decay into photons alone would pump entropy into the photon component relative to the neutrino component, in an effectively *ongoing version* of the electron-positron annihilation argument described above. The neutrino temperature $T_\nu$ (and density $\rho_\nu$) would continue to be driven down relative to $T$ (and $\rho_\gamma$) throughout the radiation-dominated era and into the matter-dominated one. In the limit $T_\nu/T \to 0$ we see from (5.36) and (5.38) that such a scenario would lead to

$$\Omega_{r,0} = \Omega_\gamma = 2.48 \times 10^{-5} h_0^{-2} \qquad \beta = 1 \, . \tag{5.41}$$

In other words, the present energy density of radiation would be lower, but it would effectively *all* be in the form of photons. Insofar as the decrease in $\Omega_{r,0}$ is precisely offset by the increase in $\beta$, these changes cancel each other out. The drop in $\Omega_{r,0}$, however, has an added consequence which is not cancelled: it pushes $t_{eq}$ farther into the past, increasing the length of time over which decaying dark energy has been contributing to the background. This raises the latter's intensity, particularly at longer wavelengths. The

effect can be significant, and we will return to this possibility at the end
of this section. For the most part, however, we will stay with the values of
$\Omega_{r,0}$ and $\beta$ given by Eqs. (5.39) and (5.40).

Armed with a definition for vacuum luminosity, Eq. (5.32), and a value
for $\beta$, Eq. (5.40), we are in a position to calculate the luminosity of decaying dark energy. Noting that $\dot{V} = 3(R/R_0)^3(\dot{R}/R)V_0$ and substituting
Eqs. (5.22) and (5.34) into (5.32), we find that

$$
L_v(t) = \mathcal{L}_{v,0} V_0 \times
\begin{cases}
\left(\dfrac{t}{t_0}\right)^{-(5-8x_m)/3} \\[2ex]
\left[\dfrac{\cosh(t/\tau_0)}{\cosh(t_0/\tau_0)}\right]\left[\dfrac{\sinh(t/\tau_0)}{\sinh(t_0/\tau_0)}\right]^{-(5-8x_m)/3}
\end{cases}
\qquad (5.42)
$$

The first of these solutions corresponds to models with $\Omega_{m,0} = 1$, while the
the second holds for the general case ($0 < \Omega_{m,0} < 1$). Both results reduce
at the present time $t = t_0$ to

$$
L_{v,0} = \mathcal{L}_{v,0} V_0 , \qquad (5.43)
$$

where $\mathcal{L}_{v,0}$ is the *comoving luminosity density of decaying dark energy*:

$$
\mathcal{L}_{v,0} = \frac{9c^2 H_0^3 \, \Omega_{r,0} \beta \, x_m}{8\pi G(1 - x_m)}
$$

$$
= 4.1 \times 10^{-30} h_0 \text{ erg s}^{-1} \text{ cm}^{-3} \left(\frac{x_m}{1 - x_m}\right) . \qquad (5.44)
$$

Numerically, we find for example that

$$
\mathcal{L}_{v,0} =
\begin{cases}
3.1 \times 10^{-31} h_0 \text{ erg s}^{-1} \text{ cm}^{-3} & (x_m = 0.07) \\
1.4 \times 10^{-30} h_0 \text{ erg s}^{-1} \text{ cm}^{-3} & (x_m = 0.25) .
\end{cases}
\qquad (5.45)
$$

In principle, dark-energy decay can produce a background 10 or even 50
times more luminous than that of galaxies, as given by (2.18). Raising
the value of the branching ratio $\beta$ to 1 instead of 0.595 does not affect
these results, since this must be accompanied by a proportionate drop in
the value of $\Omega_{r,0}$ as argued above. The numbers in (5.45) do go up if
we replace the relativistic definition (5.32) of vacuum luminosity with the
thermodynamical one (5.29); but the change is modest, raising $\mathcal{L}_{v,0}$ by no
more than a factor of 1.2 (for $x_m = 0.07$). The primary reason for the high
luminosity of the decaying vacuum lies in the fact that it converts nearly
60% of its energy density into photons. By comparison, less than 1% of the
rest energy of ordinary luminous matter has gone into photons so far in the
history of the Universe.

## 5.6 Bolometric intensity

We showed in Chap. 2 that the bolometric intensity of an arbitrary distribution of sources with comoving number density $n(t)$ and luminosity $L(t)$ could be expressed as an integral over time by (2.10). Let us apply this result here to regions of decaying dark energy, for which $n_v(t)$ and $L_v(t)$ are given by (5.28) and (5.42) respectively. Putting these equations into (2.10) along with (5.21) for the scale factor, we find that

$$
Q = c\mathcal{L}_{v,0} \times \begin{cases} \displaystyle\int_{t_{eq}}^{t_0} \left(\frac{t}{t_0}\right)^{-(1-8x_m)/3} dt \\ \displaystyle\int_{t_{eq}}^{t_0} \left[\frac{\cosh(t/\tau_0)}{\cosh(t_0/\tau_0)}\right]\left[\frac{\sinh(t/\tau_0)}{\sinh(t_0/\tau_0)}\right]^{-(1-8x_m)/3} dt \,. \end{cases} \tag{5.46}
$$

The first of these integrals corresponds to models with $\Omega_{m,0} = 1$ while the second holds for the general case $(0 < \Omega_{m,0} < 1)$. The latter may be simplified with a change of variable to $y \equiv [\sinh(t/\tau_0)]^{8x_m/3}$. Using the facts that $\sinh(t_0/\tau_0) = \sqrt{(1 - \Omega_{m,0})/\Omega_{m,0}}$ and $\cosh(t_0/\tau_0) = 1/\sqrt{\Omega_{m,0}}$ along with the definition (5.27) of $t_{eq}$, both integrals reduce to the same formula:

$$
Q = Q_v\left[1 - \left(\frac{\Omega_{r,0}}{\Omega_{m,0}}\right)^{4x_m/(1-4x_m)}\right] \,. \tag{5.47}
$$

Here $Q_v$ is found with the help of (5.44) as

$$
Q_v \equiv \frac{c\mathcal{L}_{v,0}}{4H_0\,x_m} = \frac{9c^3H_0^2\Omega_{r,0}\beta}{32\pi G(1 - x_m)} = \frac{0.0094 \text{ erg s}^{-1} \text{ cm}^{-2}}{(1 - x_m)} \,. \tag{5.48}
$$

There are several points to note about this result. First, it does not depend on $V_0$, as expected since this parameter was introduced for convenience and is not physically significant (the vacuum decays uniformly throughout space). There is also no dependence on the uncertainty $h_0$ in Hubble's constant, since the two factors of $h_0$ in $H_0^2$ are cancelled out by those in $\Omega_{r,0}$. In the limit $x_m \to 0$ one sees that $Q \to 0$ as expected. In the opposite limit where $x_m \to \frac{1}{4}$, decaying dark energy attains a maximum possible bolometric intensity of $Q \to Q_v = 0.013 \text{ erg s}^{-1} \text{ cm}^{-2}$. This is 50 times the bolometric intensity due to galaxies, as given by (2.19).

The matter density $\Omega_{m,0}$ enters only weakly into this result, and plays no role at all in the limit $x_m \to \frac{1}{4}$. Based on our experience with the EBL due to galaxies, we might have expected that $Q$ would rise significantly in models with smaller values of $\Omega_{m,0}$ since these have longer ages, giving more time for the Universe to fill up with light. What is happening here, however, is that the larger values of $t_0$ are offset by larger values of $t_{eq}$ (which follow from the fact that smaller values of $\Omega_{m,0}$ imply smaller ratios of $\Omega_{m,0}/\Omega_{r,0}$). This removes contributions from the early matter-dominated era and thereby *reduces* the value of $Q$. In the limit $x_m \to \frac{1}{4}$ these two effects cancel each other out. For smaller values of $x_m$, the $t_{eq}$-effect proves to be the stronger of the two, and one finds an overall decrease in $Q$ for these cases. With $x_m = 0.07$, for instance, the value of $Q$ drops by 2% when moving from the EdS model to $\Lambda$CDM, and by another 6% when moving from $\Lambda$CDM to $\Lambda$BDM.

## 5.7 Spectral energy distribution

To obtain limits on the parameter $x_m$, we would like to calculate the spectral intensity of the background due to dark-energy decay, just as we did for galaxies in Chap. 3. For this we need to know the spectral energy distribution (SED) of the decay photons. As discussed in Sec. 5.1, theories in which the these photons are distributed with a *non*-thermal spectrum can be strongly constrained by observations of the CMB. We therefore assume an SED of the Planckian form (3.22):

$$F_v(\lambda, t) = \frac{C(t)/\lambda^5}{\exp\left[hc/kT(t)\lambda\right] - 1} , \qquad (5.49)$$

where $T(t)$ is the blackbody temperature. The function $C(t)$ is found as usual by normalization, Eq. (3.1). Changing the integration variable from $\lambda$ to $v = c/\lambda$, we find

$$L_v(t) = \frac{C(t)}{c^4} \int_0^\infty \frac{v^3 dv}{\exp\left[hv/kT(t)\right] - 1} = \frac{C(t)}{c^4} \left[\frac{h}{kT(t)}\right]^{-4} \Gamma(4)\,\zeta(4) . \quad (5.50)$$

Inserting our result (5.42) for $L_v(t)$ and using the facts that $\Gamma(4) = 3! = 6$ and $\zeta(4) = \pi^4/90$, we obtain for $C(t)$:

$$C(t) = \frac{15\mathcal{L}_{v,0}V_0}{\pi^4} \left[\frac{hc}{kT(t)}\right]^4 \times \begin{cases} \left(\dfrac{t}{t_0}\right)^{-(5-8x_m)/3} \\[2ex] \left[\dfrac{\cosh(t/\tau_0)}{\cosh(t_0/\tau_0)}\right]\left[\dfrac{\sinh(t/\tau_0)}{\sinh(t_0/\tau_0)}\right]^{-(5-8x_m)/3} \end{cases} . \tag{5.51}$$

Here the upper expression refers as usual to the EdS case ($\Omega_{m,0} = 1$), while the lower applies to the general case ($0 < \Omega_{m,0} < 1$). The temperature of the photons can be specified if we assume thermal equilibrium between those created by vacuum decay and those already present. Stefan's law then relates $T(t)$ to the radiation energy density $\rho_r(t)c^2$ as follows:

$$\rho_r(t)\, c^2 = \frac{4\sigma_{\text{SB}}}{c}\, [T(t)]^4 \; . \tag{5.52}$$

Putting Eq. (5.23) into this expression, we find that

$$\frac{hc}{kT(t)} = \lambda_v \times \begin{cases} \left(\dfrac{t}{t_0}\right)^{2(1-x_m)/3} & (\Omega_{m,0} = 1) \\[2ex] \left[\dfrac{\sinh(t/\tau_0)}{\sinh(t_0/\tau_0)}\right]^{2(1-x_m)/3} & (0 < \Omega_{m,0} < 1)\,, \end{cases} \tag{5.53}$$

where the constant $\lambda_v$ is given by

$$\lambda_v \equiv \left(\frac{8\pi^5 hc}{15\rho_{r,0}\, c^2}\right)^{1/4} = 0.46 \text{ cm} \left(\frac{\Omega_{r,0} h_0^2}{4.17\times 10^{-5}}\right)^{-1/4} . \tag{5.54}$$

This value of $\lambda_v$ confirms that vacuum decay will be concentrated in the microwave region near the CMB peak, as expected ($\lambda_{\text{cmb}} = 0.11$ cm). Putting (5.53) back into (5.51), we obtain

$$C(t) = \frac{15\lambda_v^4 \mathcal{L}_{v,0} V_0}{\pi^4} \times \begin{cases} \left(\dfrac{t}{t_0}\right) \\[2ex] \left[\dfrac{\cosh(t/\tau_0)}{\cosh(t_0/\tau_0)}\right]\left[\dfrac{\sinh(t/\tau_0)}{\sinh(t_0/\tau_0)}\right] . \end{cases} \tag{5.55}$$

These two expressions refer to models with $\Omega_{m,0} = 1$ and $0 < \Omega_{m,0} < 1$ respectively. This specifies the SED (5.49) of decaying dark energy.

## 5.8 Dark energy and the background light

The spectral intensity of an arbitrary distribution of sources with comoving number density $n(t)$ and an SED $F(\lambda,t)$ is expressed as an integral over time by Eq. (3.5). With Eqs. (5.21), (5.28) and (5.49), we find

$$I_\lambda(\lambda_0) = I_v(\lambda_0) \times \begin{cases} \displaystyle\int_{t_{\text{eq}}/t_0}^{1} \frac{\tau^{-1}\, d\tau}{\exp\left[\left(\dfrac{\lambda_v}{\lambda_0}\right)\tau^{-2x_m/3}\right] - 1} \\[4ex] \displaystyle\int_{t_{\text{eq}}/\tau_0}^{t_0/\tau_0} \frac{\coth\tau\, d\tau}{\exp\left[\dfrac{\lambda_v}{\lambda_0}\left(\dfrac{\sqrt{\Omega_{m,0}}\,\sinh\tau}{\sqrt{1-\Omega_{m,0}}}\right)^{-2x_m/3}\right] - 1} . \end{cases} \tag{5.56}$$

Here we have used integration variables $\tau \equiv t/t_0$ in the first case ($\Omega_{m,0} = 1$) and $\tau \equiv t/\tau_0$ in the second ($0 < \Omega_{m,0} < 1$). The dimensional content of both integrals is contained in the prefactor $I_v(\lambda_0)$, which reads

$$I_v(\lambda_0) \equiv \frac{5\mathcal{L}_{v,0}}{2\pi^5 h H_0} \left(\frac{\lambda_v}{\lambda_0}\right)^4 = 15,500 \text{ CUs} \left(\frac{x_m}{1-x_m}\right)\left(\frac{\lambda_v}{\lambda_0}\right)^4 . \quad (5.57)$$

We have divided this quantity through by photon energy $hc/\lambda_0$ so as to express the results in continuum units (CUs) as usual, where 1 CU $\equiv$ 1 photon s$^{-1}$ cm$^{-2}$ Å$^{-1}$ ster$^{-1}$. We use CUs throughout this book, for the sake of uniformity as well as the fact that these units carry several advantages from the theoretical point of view (Sec. 3.2). The reader who consults the literature, however, will soon find that each part of the electromagnetic spectrum has its own "dialect" of preferred units. In the microwave region, intensities are commonly reported in terms of $\nu I_\nu$, the integral of flux per unit frequency *over* frequency, and usually expressed in units of nW m$^{-2}$ ster$^{-1}$ = $10^{-6}$ erg s$^{-1}$ cm$^{-2}$ ster$^{-1}$. To translate a given value of $\nu I_\nu$ (in these units) into CUs, we need only multiply by a factor of $10^{-6}/(hc)$ = 50.34 erg$^{-1}$ Å$^{-1}$. The Jansky (Jy) is also often encountered, with 1 Jy = $10^{-23}$ erg s$^{-1}$ cm$^{-2}$ Hz$^{-1}$. To convert a given value of $\nu I_\nu$ from Jy ster$^{-1}$ into CUs, we multiply by a factor of $10^{-23}/h\lambda = (1509$ Hz erg$^{-1})/\lambda$ with $\lambda$ in Å.

Eq. (5.56) gives the combined intensity of decay photons which have been emitted at many wavelengths and redshifted by various amounts, but reach us in a waveband centered on $\lambda_0$. The arbitrary volume $V_0$ has dropped out of the integral as expected, and this result is also independent of the uncertainty $h_0$ in Hubble's constant since there is a factor of $h_0$ in both $\mathcal{L}_{v,0}$ and $H_0$. Results are plotted in Fig. 5.2. over the waveband 0.01-1 cm, together with existing observational data in this part of the spectrum. The most celebrated of these is the COBE detection of the CMB [227], which we have shown as a heavy solid line (F96). The experimental uncertainties in this measurement are far smaller than the thickness of the line. The other observational limits shown in Fig. 5.2 have been obtained in the far infrared (FIR) region, also from analysis of data from the COBE satellite. These are indicated with heavy dotted lines (F98 [233]) and open triangles (H98 [64] and L00 [234]).

Fig. 5.2(a) shows the spectral intensity of background radiation from vacuum decay under our standard assumptions, including the relativistic definition (5.32) of vacuum luminosity and the values of $\Omega_{r,0}$ and $\beta$ given by (5.39) and (5.40) respectively. Five groups of curves are shown, corresponding to values of $x_m$ between $3 \times 10^{-5}$ and the theoretical maximum of

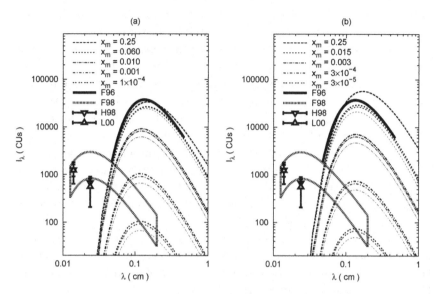

Fig. 5.2    The spectral intensity of background radiation due to the decaying vacuum for various values of $x_m$, compared with observational data in the microwave region (heavy solid line) and far infrared (heavy dotted line and squares). For each value of $x_m$ there are three curves representing cosmologies with $\Omega_{m,0} = 1$ (heaviest lines), $\Omega_{m,0} = 0.3$ (medium-weight lines) and $\Omega_{m,0} = 0.03$ (lightest lines). Panel (a) assumes $L = L_{rel}$ and $\beta = 0.595$, while panel (b) assumes $L = L_{th}$ and $\beta = 1$.

0.25. For each value of $x_m$ three curves are plotted: one each for the EdS, $\Lambda$CDM and $\Lambda$BDM cosmologies. As noted above in connection with the bolometric intensity $Q$, the choice of cosmological model is less important in determining the background due to vacuum decay than the background due to galaxies. In fact, the intensities here are actually slightly *lower* in vacuum-dominated models. The reason for this, as before, is that these models have smaller values of $\Omega_{m,0}/\Omega_{r,0}$ and hence larger values of $t_{eq}$, reducing the size of contributions from the early matter-dominated era when $L_v$ is large.

In Fig. 5.2(b), we have exchanged the relativistic definition of vacuum luminosity for the thermodynamical one (5.29), and set $\beta = 1$ instead of 0.595. As discussed in Sec. 5.5, the increase in $\beta$ is partly offset by a drop in $\Omega_{r,0}$. There is a net increase in intensity, however, because smaller values of $\Omega_{r,0}$ push $t_{eq}$ back into the past, leading to additional contributions from the early matter-dominated era. These contributions particularly push up the long-wavelength part of the spectrum in Fig. 5.2(b) relative to Fig. 5.2(a), as seen most clearly in the case $x_m = 0.25$. Overall, intensities in Fig. 5.2(b) are higher than those in Fig. 5.2(a) by about a factor of four.

These figures show that *the decaying-vacuum hypothesis is strongly constrained by observations of the microwave background.* The parameter $x_m$ cannot be larger than 0.06 or the intensity of the decaying vacuum would exceed that of the CMB itself under the most conservative assumptions, as represented by Fig. 5.2(a). This limit tightens to $x_m \leqslant 0.015$ if different assumptions are made about the luminosity of the vacuum, as shown by Fig. 5.2(b). These numbers are comparable to the limit of $x \leqslant 0.07$ obtained from entropy conservation under the assumption that $x = x_r = x_m$ [211]. And insofar as the CMB is usually attributed entirely to relic radiation from the big bang, the real limit on $x_m$ is probably several orders of magnitude smaller than this.

With these upper bounds on $x_m$, we can finally inquire about the potential of the decaying vacuum as a dark-energy candidate. Since its density is given by (5.17) as a fraction $x/(1 - x)$ of that of radiation, we infer that its present density parameter ($\Omega_{v,0}$) satisfies:

$$\Omega_{v,0} = \left(\frac{x_m}{1 - x_m}\right) \Omega_{r,0} \leqslant \begin{cases} 7 \times 10^{-6} & \text{(a)} \\ 1 \times 10^{-6} & \text{(b)} . \end{cases} \tag{5.58}$$

Here, (a) and (b) refer to the scenarios represented by Figs. 5.2(a) and 5.2(b), with the corresponding values of $\Omega_{r,0}$ as defined by Eqs. (5.39) and (5.41) respectively. We have assumed that $h_0 \geqslant 0.6$ as usual. It is clear from the limits (5.58) that a decaying vacuum of the kind we have considered here does not contribute significantly to the density of the dark energy.

It should be recalled, however, that there are good reasons from quantum theory for expecting some kind of instability for the vacuum in a universe which progressively cools. (Equivalently, there are good reasons for believing that the cosmological "constant" is not.) Our conclusion is that if the vacuum decays, it either does so very slowly, or in a manner that does not upset the isotropy of the CMB.

# Chapter 6

# The Infrared and Visible Backgrounds

## 6.1 Decaying axions

Axions are hypothetical particles whose existence would explain what is otherwise a puzzling feature of quantum chromodynamics (QCD), the leading theory of strong interactions. QCD contains a dimensionless free parameter ($\Theta$) whose value must be "unnaturally" small in order for the theory not to violate a combination of charge conservation and mirror-symmetry known as charge parity or CP. Upper limits on the electric dipole moment of the neutron currently constrain the value of $\Theta$ to be less than about $10^{-9}$. The strong CP problem is the question: "Why is $\Theta$ so small?" This is reminiscent of the cosmological-constant problem (Sec. 4.5), though less severe by many orders of magnitude. Proposed solutions have similarly focused on making $\Theta$, like $\Lambda$, a dynamical variable whose value could have been driven toward zero in the early Universe. In the most widely-accepted scenario, due to R. Peccei and H. Quinn in 1977 [235], this is accomplished by the spontaneous breaking of a new global symmetry (now called PQ symmetry) at energy scales $f_{\mathrm{PQ}}$. As shown by S. Weinberg [236] and F. Wilczek [237] in 1978, the symmetry-breaking gives rise to a new particle which eventually acquires a rest energy $m_a c^2 \propto f_{\mathrm{PQ}}^{-1}$. This particle is the axion ($a$).

Axions, if they exist, meet all the requirements of a successful CDM candidate (Sec. 4.3): they interact weakly with the baryons, leptons and photons of the standard model; they are cold (i.e. non-relativistic during the time when structure begins to form); and they are capable of providing some or even all of the CDM density which is thought to be required, $\Omega_{\mathrm{cdm}} \sim 0.3$. A fourth property, and the one which is of most interest to us here, is that *axions decay generically into photon pairs*. The importance of this process depends on two things: the axion's rest mass $m_a$ and its coupling strength $g_{a\gamma\gamma}$. The PQ symmetry-breaking energy scale $f_{\text{PQ}}$ is not constrained by the theory, and reasonable values for this parameter are such that $m_a c^2$ might in principle lie anywhere between $10^{-12}$ eV and 1 MeV [129]. This broad range of theoretical possibilities has been whittled down by an impressive array of cosmological, astrophysical and laboratory-based tests. In summarizing these, it is useful to distinguish between axions with rest energies above and below a "critical" rest energy $m_{a,\text{crit}} c^2 \sim 3 \times 10^{-2}$ eV.

Axions whose rest energies lie *below* $m_{a,\text{crit}} c^2$ arise primarily via processes known as vacuum misalignment [238–240] and axionic string decay [241]. These are non-thermal mechanisms, meaning that the axions produced in this way were never in thermal equilibrium with the primordial plasma. Their present density would be at least [242]

$$\Omega_a \approx \left( \frac{m_a c^2}{4 \times 10^{-6} \text{ eV}} \right)^{-7/6} h_0^{-2} \,. \tag{6.1}$$

This number is currently under debate, and may go up by an order of magnitude or more if string effects play an important role [243]. If we require that axions not provide *too much* CDM ($\Omega_{\text{cdm}} \leqslant 0.6$), then (6.1) implies a lower limit on the axion rest energy:

$$m_a c^2 \gtrsim 7 \times 10^{-6} \,. \tag{6.2}$$

This neatly eliminates the lower third of the theoretical axion mass-window. Upper limits on $m_a$ for non-thermal axions have come from astrophysics. Prime among these is the fact that the weak couplings of axions to baryons, leptons and photons allow them to stream freely from stellar cores, carrying energy with them. If they are massive enough, such axions could in principle cool the core of the Sun, alter the helium-burning phase in red-giant stars, and shorten the duration of the neutrino burst from supernovae such as SN1987a. The last of these effects is particularly sensitive and requires [244, 245]:

$$m_a c^2 \lesssim 6 \times 10^{-3} \text{ eV} \,. \tag{6.3}$$

Axions with $10^{-5} \lesssim m_a c^2 \lesssim 10^{-2}$ thus remain compatible with both cosmological and astrophysical limits, and could provide much or all of the CDM. It may be possible to detect these particles in the laboratory by enhancing their conversion into photons with strong magnetic fields, as suggested by P. Sikivie in 1983 [246]. Experimental search programs based on this principle are now in operation in California [247], Tokyo [248], the Sierra Grande mountains in Argentina (SOLAX [249]), the Spanish Pyrenees (COSME [250])

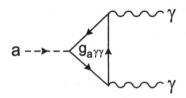

Fig. 6.1 The Feynman diagram corresponding to the decay of the axion ($a$) into two photons ($\gamma$) with coupling strength $g_{a\gamma\gamma}$.

and CERN in Switzerland (CAST [251]). Exclusion plots from these experiments are beginning to restrict theoretically-favored regions of the phase space defined by $m_a$ and $g_{a\gamma\gamma}$. It has also been suggested that the Universe itself might behave like a giant axion cavity detector, since it is threaded by intergalactic magnetic fields. These could stimulate the reverse process, converting photons into a different kind of "ultralight" axion and possibly explaining the apparent dimming of high-redshift Type Ia supernovae without vacuum energy [252].

Promising as they are, we will not consider non-thermal axions (sometimes known as "invisible axions") further in this section. This is because they decay too slowly to leave any trace in the extragalactic background light. Axions decay into photon pairs ($a \rightarrow \gamma + \gamma$) via a loop diagram, as illustrated in Fig. 6.1. The *decay lifetime* of this process is [129]

$$\tau_a = (6.8 \times 10^{24} \text{ s}) \, m_1^{-5} \, \zeta^{-2} \,. \tag{6.4}$$

Here $m_1 \equiv m_a c^2/(1 \text{ eV})$ is the axion rest energy in units of eV, and $\zeta$ is a constant which is proportional to the coupling strength $g_{a\gamma\gamma}$ [253]. For our purposes, it is sufficient to treat $\zeta$ as a free parameter which depends on the details of the axion theory chosen. Its value has been normalized in Eq. (6.4), so that $\zeta = 1$ in the simplest grand-unified theories (GUTs) for the strong and electroweak interactions. This could drop to $\zeta = 0.07$ in other theories, however [254], strongly suppressing the two-photon decay channel. In principle $\zeta$ could even vanish altogether, corresponding to a radiatively *stable* axion, although this would require an unlikely cancellation of terms. We will consider values in the range $0.07 \leqslant \zeta \leqslant 1$ in what follows. For these values of $\zeta$, and with $m_1 \lesssim 6 \times 10^{-3}$ as given by (6.3), Eq. (6.4) shows that axions decay on timescales $\tau_a \gtrsim 9 \times 10^{35}$ s. This is much longer than the age of the Universe, so that such particles would truly be "invisible."

We therefore shift our attention to axions with rest energies *above* $m_{a,\text{crit}}c^2$. M.S. Turner showed in 1987 [255] that the vast majority of these would have arisen in the early Universe via thermal mechanisms such as Primakoff scattering and photo-production. The Boltzmann equation can be solved to give their present comoving number density as $n_a = (830/g_{*F}) \text{ cm}^{-3}$ [253], where $g_{*F} \approx 15$ counts the number of relativistic degrees of freedom left in the plasma at the time when axions "froze out" of equilibrium. The present density parameter $\Omega_a = n_a m_a/\rho_{\text{crit},0}$ of thermal axions is thus

$$\Omega_a = 5.2 \times 10^{-3} h_0^{-2} m_1 \ . \tag{6.5}$$

Whether or not this is dynamically significant depends on the axion rest mass. The duration of the neutrino burst from SN1987a again imposes a powerful constraint on $m_a c^2$. This time, however, it is a *lower* bound, because axions in this range of rest energies are massive enough to interact with nucleons in the supernova core and can no longer stream out freely. Instead, they are trapped in the core and radiate only from an "axiosphere" rather than the entire volume of the star. Axions with sufficiently large $m_a c^2$ are trapped so strongly that they no longer interfere with the luminosity of the neutrino burst, leading to the lower limit [256]

$$m_a c^2 \gtrsim 2.2 \text{ eV} \ . \tag{6.6}$$

Astrophysics also provides strong upper bounds on $m_a c^2$ for thermal axions. These depend critically on whether or not axions couple only to hadrons, or to other particles as well. An early class of *hadronic* axions (which couple only to hadrons) was studied by J.E. Kim [257] and M.A. Shifman, A.I. Vainshtein and V.I. Zakharov [258]. These particles are often termed KSVZ axions. Another widely-discussed model in which axions couple to charged leptons as well as nucleons and photons has been discussed by A.R. Zhitnitsky [259] and M. Dine, W. Fischler and M. Srednicki [260]. These particles are known as DFSZ axions. The extra lepton coupling of these DFSZ axions allows them to carry so much energy out of the cores of red-giant stars that helium ignition is seriously disrupted unless $m_a c^2 \lesssim 9 \times 10^{-3}$ eV [261]. Since this upper limit is inconsistent with the lower limit (6.6), thermal DFSZ axions are excluded. For KSVZ or hadronic axions, red giants impose a weaker bound [262]:

$$m_a c^2 \lesssim 0.7 \zeta^{-1} \text{ eV} \ . \tag{6.7}$$

This is consistent with the lower limit (6.6) for realistic values of the parameter $\zeta$. For the simplest hadronic axion models with $\zeta \geqslant 0.07$, for instance,

Eq. (6.7) translates into an upper limit $m_a c^2 \lesssim 10$ eV. It has been argued that axions with $m_a c^2 \gtrsim 10$ eV can be ruled out in any case, because they would interact strongly enough with baryons to produce a detectable signal in existing Čerenkov detectors [263].

For thermally-produced hadronic axions, then, there remains a window of opportunity in the multi-eV range with $2 \lesssim m_1 \lesssim 10$. Eq. (6.5) shows that these particles would contribute a total density of about $0.03 \lesssim \Omega_a \lesssim 0.15$, where we take $0.6 \leqslant h_0 \leqslant 0.9$ as usual. They would not be able to provide the entire density of dark matter in the $\Lambda$CDM model ($\Omega_{m,0} = 0.3$), but they could suffice in low-density models midway between $\Lambda$CDM and $\Lambda$BDM (Table 3.1). Since such models remain compatible with most current observational data (Chap. 4), it is worth proceeding to see how these multi-eV axions can be further constrained by their contributions to the EBL.

## 6.2 Axion halos

Thermal axions are not as cold as their non-thermal cousins, but will still be found primarily inside gravitational potential wells such as those of galaxies and galaxy clusters [255]. We need not be too specific about the fraction which have settled into galaxies as opposed to larger systems, because we will be concerned primarily with their *combined* contributions to the diffuse background. (Distribution could become an issue if extinction due to dust or gas played a strong role inside the bound regions, but this is not likely to be important for the photon energies under consideration here.) These axion halos provide us with a convenient starting-point as cosmological light sources, analogous to the galaxies and vacuum source regions of previous sections. Let us take the axions to be cold enough that their fractional contribution ($M_h$) to the total mass of each halo ($M_{\text{tot}}$) is the same as their fractional contribution to the cosmological matter density, $M_h/M_{\text{tot}} = \Omega_a/\Omega_{m,0} = \Omega_a/(\Omega_a + \Omega_{\text{bar}})$. Then the mass $M_h$ of axions in each halo is

$$M_h = M_{\text{tot}} \left( 1 + \frac{\Omega_{\text{bar}}}{\Omega_a} \right)^{-1} . \tag{6.8}$$

Here we have made the minimal assumption that axions constitute *all* the nonbaryonic dark matter. If the halos are distributed with a mean comoving number density $n_0$, then the cosmological density of bound axions is $\Omega_{a,\text{bound}} = (n_0 M_h)/\rho_{\text{crit},0} = (n_0 M_{\text{tot}}/\rho_{\text{crit},0})(1 + \Omega_{\text{bar}}/\Omega_a)^{-1}$. Equating

$\Omega_{a,\text{bound}}$ to $\Omega_a$, as given by (6.5), fixes the total mass:

$$M_{\text{tot}} = \frac{\Omega_a \rho_{\text{crit},0}}{n_0} \left(1 + \frac{\Omega_{\text{bar}}}{\Omega_a}\right) . \tag{6.9}$$

The comoving number density of galaxies at $z = 0$ is [232]

$$n_0 = 0.010\, h_0^3 \text{ Mpc}^{-3} . \tag{6.10}$$

Using this together with (6.5) for $\Omega_a$, and setting $\Omega_{\text{bar}} \approx 0.016 h_0^{-2}$ from Sec. 4.2, we find from (6.9) that

$$M_{\text{tot}} = \begin{cases} 9 \times 10^{11} M_\odot h_0^{-3} & (m_1 = 3) \\ 1 \times 10^{12} M_\odot h_0^{-3} & (m_1 = 5) \\ 2 \times 10^{12} M_\odot h_0^{-3} & (m_1 = 8) \end{cases} . \tag{6.11}$$

Let us compare these numbers with dynamical data on the mass of the Milky Way using the motions of Galactic satellites. These assume a Jaffe profile [264] for halo density:

$$\rho_{\text{tot}}(r) = \frac{v_c^2}{4\pi G\, r^2} \frac{r_j^2}{(r + r_j)^2} . \tag{6.12}$$

Here $v_c$ is the circular velocity, $r_j$ the Jaffe radius, and $r$ the radial distance from the center of the Galaxy. The data imply that $v_c = 220 \pm 30$ km s$^{-1}$ and $r_j = 180 \pm 60$ kpc [110]. Integrating over $r$ from zero to infinity gives

$$M_{\text{tot}} = \frac{v_c^2 r_j}{G} = (2 \pm 1) \times 10^{12} M_\odot . \tag{6.13}$$

This is consistent with (6.11) for most values of $m_1$ and $h_0$. So axions of this type could in principle make up all the dark matter which is required on Galactic scales.

Putting (6.9) into (6.8) gives the mass of an axion halo as

$$M_h = \frac{\Omega_a \rho_{\text{crit},0}}{n_0} . \tag{6.14}$$

This could also have been derived as the mass of a region of space of comoving volume $V_0 = n_0^{-1}$ filled with homogeneously-distributed axions of mean density $\rho_a = \Omega_a \rho_{\text{crit},0}$. (This is the approach that we adopted in defining vacuum regions in Sec. 5.5.)

To obtain the halo luminosity, we sum up the rest energies of all the decaying axions in the halo and divide by the decay lifetime (6.4):

$$L_h = \frac{M_h c^2}{\tau_a} . \tag{6.15}$$

Inserting Eqs. (6.4) and (6.14), we find

$$L_h = (3.8 \times 10^{40} \text{ erg s}^{-1}) h_0^{-3} \zeta^2 m_1^6$$

$$= \begin{cases} 7 \times 10^9 L_\odot \, h_0^{-3} \zeta^2 & (m_1 = 3) \\ 2 \times 10^{11} L_\odot \, h_0^{-3} \zeta^2 & (m_1 = 5) \\ 3 \times 10^{12} L_\odot \, h_0^{-3} \zeta^2 & (m_1 = 8) \end{cases} . \tag{6.16}$$

The luminosities of the galaxies themselves are of order $L_0 = \mathcal{L}_0/n_0 = 2 \times 10^{10} h_0^{-2} L_\odot$, where we have used (2.18) for $\mathcal{L}_0$. Thus axion halos could in principle outshine their host galaxies, unless axions are either very light ($m_1 \lesssim 3$) or weakly-coupled ($\zeta < 1$). This already suggests that they will be strongly constrained by observations of EBL intensity.

## 6.3 Bolometric intensity

Substituting the halo comoving number density (6.10) and luminosity (6.16) into Eq. (2.13), we find that the combined intensity of decaying axions at all wavelengths is given by

$$Q = Q_a \int_0^{z_f} \frac{dz}{(1+z)^2 \tilde{H}(z)} . \tag{6.17}$$

Here the dimensional content of the integral is contained in the prefactor $Q_a$, which takes the following numerical values:

$$Q_a = \frac{\Omega_a \rho_{\text{crit},0} \, c^3}{H_0 \tau_a} = (1.2 \times 10^{-7} \text{ erg s}^{-1} \text{ cm}^{-2}) h_0^{-3} \zeta^2 m_1^6 \tag{6.18}$$

$$= \begin{cases} 9 \times 10^{-5} \text{ erg s}^{-1} \text{ cm}^{-2}) h_0^{-1} \zeta^2 & (m_1 = 3) \\ 2 \times 10^{-3} \text{ erg s}^{-1} \text{ cm}^{-2}) h_0^{-1} \zeta^2 & (m_1 = 5) \\ 3 \times 10^{-2} \text{ erg s}^{-1} \text{ cm}^{-2}) h_0^{-1} \zeta^2 & (m_1 = 8) \end{cases} .$$

There are three things to note about this quantity. First, it is comparable in magnitude to the *observed* EBL due to galaxies, $Q_* \approx 3 \times 10^{-4} \text{ erg s}^{-1} \text{ cm}^{-2}$ (Chap. 2). Second, unlike $Q_*$ for galaxies or $Q_v$ for decaying vacuum energy, $Q_a$ depends explicitly on the uncertainty $h_0$ in Hubble's constant. Physically, this reflects the fact that the axion density $\rho_a = \Omega_a \rho_{\text{crit},0}$ in the numerator of (6.18) comes to us from the Boltzmann equation and is independent of $h_0$, whereas the density of luminous matter in galaxies comes from the luminosity density $\mathcal{L}_0$ (which is proportional to $h_0$ and so cancels the $h_0$-dependence in $H_0$). The third thing to note about $Q_a$ is that it is independent of $n_0$. This is because the collective contribution of decaying axions to the diffuse background is determined by their mean density $\Omega_a$, and does not depend on how they are distributed in space.

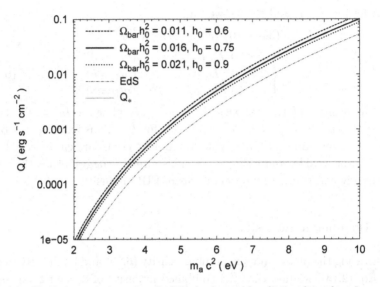

Fig. 6.2   The bolometric intensity $Q$ of the background radiation from decaying axions as a function of their rest masses $m_a$. The faint dash-dotted line shows the equivalent intensity in an EdS model (in which axions alone cannot provide the required CDM). The dotted horizontal line indicates the characteristic bolometric intensity $(Q_*)$ of the observed EBL.

To evaluate (6.17) we need to specify the cosmological model. If we assume a spatially flat Universe, as increasingly suggested by the data (Chap. 4), then Hubble's parameter (2.31) reads

$$\tilde{H}(z) = \left[\Omega_{m,0}(1 + z)^3 + 1 - \Omega_{m,0}\right]^{1/2} . \tag{6.19}$$

Here, we are taking the most economical approach and requiring axions to make up all the cold dark matter so that $\Omega_{m,0} = \Omega_a + \Omega_{bar}$. Putting this into Eq. (6.17) along with (6.18) for $Q_a$, we obtain the plots of $Q(m_1)$ shown in Fig. 6.2 for $\zeta = 1$. The three heavy lines in this plot show the range of intensities obtained by varying $h_0$ and $\Omega_{bar}h_0^2$ within the ranges $0.6 \leqslant h_0 \leqslant 0.9$ and $0.011 \leqslant \Omega_{bar}h_0^2 \leqslant 0.021$ respectively. We have set $z_f = 30$, since axions were presumably decaying long before they became bound to galaxies. (Results are however fairly insensitive to this choice.)

The axion-decay background is faintest for the largest values of $h_0$, as expected from the fact that $Q_a \propto h_0^{-1}$. This is partly offset by the fact that larger values of $h_0$ also lead to a drop in $\Omega_{m,0}$, extending the age of the Universe and hence the length of time over which axions have been contributing to the background. (Smaller values of $\Omega_{bar}$ raise the intensity slightly for the same reason.) Overall, Fig. 6.2 shows that axions with $\zeta = 1$ and $m_a c^2 \gtrsim 3.5$ eV are capable of producing a background brighter than that from the galaxies themselves.

## 6.4 Axions and the background light

To go further and compare our predictions with observational data, we would like to calculate the intensity of axionic contributions to the EBL as a function of wavelength. The first step, as usual, is to specify the spectral energy distribution or SED of the decay photons in the rest frame. Each axion decays into two photons of energy $\frac{1}{2} m_a c^2$ (Fig. 6.1), so that the decay photons are emitted at or near a peak wavelength

$$\lambda_a = \frac{2hc}{m_a c^2} = \frac{24,800 \text{ Å}}{m_1} . \tag{6.20}$$

Since $2 \lesssim m_1 \lesssim 10$, the value of this parameter tells us that we will be most interested in the *infrared and optical* bands (roughly 4000–40,000 Å). We can model the decay spectrum with a Gaussian SED as in (3.19):

$$F(\lambda) = \frac{L_h}{\sqrt{2\pi}\,\sigma_\lambda} \exp\left[ -\frac{1}{2}\left( \frac{\lambda - \lambda_a}{\sigma_\lambda} \right)^2 \right] . \tag{6.21}$$

The width $\sigma_\lambda$ of the curve is related to the dispersion $v_c$ of the bound axions [266]. For the Milky Way, $v_c = 220$ km s$^{-1}$, so that $\sigma_\lambda = 2(v_c/c)\lambda_a \approx 40$ Å/$m_1$. For axions bound in galaxy clusters, $v_c$ rises to as much as 1300 km s$^{-1}$ [253], implying that $\sigma_\lambda \approx 220$ Å/$m_1$. Let us parametrize $\sigma_\lambda$ in terms of a dimensionless quantity $\sigma_{50} \equiv \sigma_\lambda/(50 \text{ Å}/m_1)$ so that

$$\sigma_\lambda = (50 \text{ Å}/m_1)\,\sigma_{50} . \tag{6.22}$$

With the SED $F(\lambda)$ thus specified along with Hubble's parameter (6.19), the spectral intensity of the background radiation produced by axion decays is given by (3.6) as

$$I_\lambda(\lambda_0) = I_a \int\limits_0^{z_f} \frac{\exp\left\{ -\frac{1}{2}\left[ \frac{\lambda_0/(1+z) - \lambda_a}{\sigma_\lambda} \right]^2 \right\} dz}{(1+z)^3 [\Omega_{m,0}(1+z)^3 + 1 - \Omega_{m,0}]^{1/2}} . \tag{6.23}$$

The dimensional prefactor in this case reads

$$I_a = \frac{\Omega_a \rho_{\text{crit},0}\, c^2}{\sqrt{32\pi^3}\, h\, H_0\, \tau_a} \left( \frac{\lambda_0}{\sigma_\lambda} \right)$$

$$= (95 \text{ CUs})\, h_0^{-1} \zeta^2 m_1^7 \sigma_{50}^{-1} \left( \frac{\lambda_0}{24,800 \text{ Å}} \right) . \tag{6.24}$$

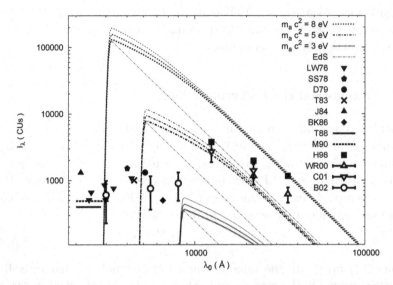

Fig. 6.3    The spectral intensity $I_\lambda$ of the background radiation from decaying axions as a function of observed wavelength $\lambda_0$. The four curves for each value of $m_a$ (labelled) correspond to upper, median and lower limits on $h_0$ and $\Omega_{bar}$ together with the equivalent intensity for the EdS model (as in Fig. 6.2). Also shown are observational upper limits (solid symbols and heavy lines) and reported detections (empty symbols) over this waveband.

Here, we have divided through by the photon energy $hc/\lambda_0$ to put results into continuum units or CUs as usual (Sec. 3.2). The number density in (3.6) cancels out the factor of $1/n_0$ in luminosity (6.16), so that results are independent of axion distribution, as expected.

Evaluating Eq. (6.23) over 2000 Å$\leqslant \lambda_0 \leqslant$ 20,000 Å with $\zeta = 1$ and $z_f = 30$, we obtain the plots of $I_\lambda(\lambda_0)$ shown in Fig. 6.3. Three groups of curves are shown, corresponding to $m_a c^2 = 3$ eV, 5 eV and 8 eV. For each value of $m_a$ there are four curves; these assume $(h_0, \Omega_{bar}h_0^2) = (0.6, 0.011), (0.75, 0.016)$ and $(0.9, 0.021)$ respectively, with the fourth (faint dash-dotted) curve representing the equivalent intensity in an EdS universe (as in Fig. 6.2). Also plotted are many of the reported observational constraints on EBL intensity in this waveband. Most have been encountered already in Chap. 3. They include data from the OAO-2 satellite (LW76 [57]), several ground-based telescope observations (SS78 [58], D79 [59], BK86 [60]), the Pioneer 10 spacecraft (T83 [55]), sounding rockets (J84 [61], T88 [62]), the Space Shuttle-borne Hopkins UVX experiment (M90 [63]), the DIRBE instrument aboard the COBE satellite (H98 [64], WR00 [65], C01

[66]), and a combination of ground and space-based observations using Las Campanas and the Hubble Space Telescope (HST).

Fig. 6.3 shows that 8 eV axions with $\zeta = 1$, if they existed, would produce a hundred times more background light at $\sim 3000$ Å than is actually seen. The background from 5 eV axions would similarly exceed observed levels by a factor of ten at $\sim 5000$ Å (incidentally coloring the night sky a bright shade of green). Only axions with $m_a c^2 \leqslant 3$ eV are compatible with observation if $\zeta = 1$. These results are brighter than ones obtained assuming an EdS cosmology [265], especially at wavelengths longward of the peak. This reflects the fact that the background in a low-$\Omega_{m,0}$, high-$\Omega_{\Lambda,0}$ universe like that considered here receives many more contributions from sources at high redshift.

We can obtain constraints as a smooth function of the axion rest mass as follows: since $I_\lambda \propto \zeta^{-2}$, the value of $\zeta$ required to reduce the minimum predicted axion intensity $I_{\rm th}$ below a given observational upper limit $I_{\rm obs}$ at any wavelength $\lambda_0$ in Fig. 6.3 is $\zeta \leqslant \sqrt{I_{\rm obs}/I_{\rm th}}$. The upper limit on $\zeta$ (for a given value of $m_a$) is then the smallest such value of $\zeta$; i.e. that which brings $I_{\rm th}$ down to $I_{\rm obs}$ or below at *each* wavelength $\lambda_0$. From this procedure we obtain a function which can be regarded as an upper limit on the axion rest mass $m_a$ as a function of $\zeta$ (or vice versa). Results are plotted in Fig. 6.4 (heavy solid line). This curve tells us that even in models where the axion-photon coupling is strongly suppressed and $\zeta = 0.07$, the axion cannot be more massive than

$$m_a c^2 \leqslant 8.0 \text{ eV} \qquad (\zeta = 0.07) \, . \tag{6.25}$$

In the simplest axion models with $\zeta = 1$, this limit tightens to

$$m_a c^2 \leqslant 3.7 \text{ eV} \qquad (\zeta = 1) \, . \tag{6.26}$$

As expected, these bounds are stronger than those obtained in an EdS model, for which some other CDM candidate would have to be postulated besides the axions (Fig. 6.4, faint dotted line). This is a small effect, however, because the strongest constraints tend to come from the region near the peak wavelength ($\lambda_a$), whereas the difference between matter- and vacuum-dominated models is most pronounced at wavelengths longward of the peak where the majority of the radiation originates at high redshift. Fig. 6.4 shows that cosmology in this case has the most effect over the range $0.1 \lesssim \zeta \lesssim 0.4$, where upper limits on $m_a c^2$ are weakened by about 10% in the EdS model, relative to one in which the CDM is assumed to consist only of axions.

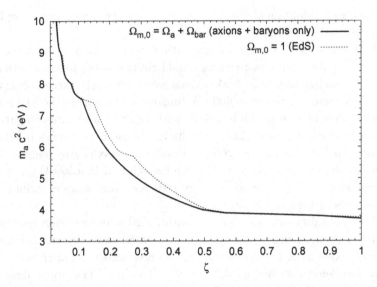

Fig. 6.4    The upper limits on the value of $m_a c^2$ as a function of the coupling strength $\zeta$ (or vice versa). These are derived by requiring that the minimum predicted axion intensity (as plotted in Fig. 6.3) be less than or equal to the observational upper limits on the intensity of the EBL.

Combining Eqs. (6.6) and (6.26), we conclude that axions in the simplest models are confined to a slender range of viable rest masses:

$$2.2 \text{ eV} \lesssim m_a c^2 \leqslant 3.7 \text{ eV} . \tag{6.27}$$

Background radiation thus complements the red-giant bound (6.7). Eq. (6.27) *closes off most, if not all of the multi-eV window for thermal axions.* The allowed range of values may be further narrowed by looking for the enhanced signal which might be expected to emanate from concentrations of bound axions associated with galaxies and clusters of galaxies, as first suggested by T.W. Kephart and T.J. Weiler in 1987 [266]. The most thorough search along these lines was reported in 1991 by M.A. Bershady, M.T. Ressell and M.S. Turner [267], who found no evidence of the expected signal from three selected clusters, further tightening the upper limit on the multi-eV axion window to 3.2 eV in the simplest models. Constraints obtained in this way for *non*-thermal axions would be considerably weaker, as noted by several workers [266, 268]; but this does not affect our results, since axions in the range of rest masses considered here are overwhelmingly thermal ones. Similarly, "invisible" axions with rest masses near the upper limit given by Eq. (6.3) might give rise to detectable microwave signals from nearby mass concentrations such as the Local Group of galaxies. This is the premise for a recent search carried out by B.D. Blout *et al.* [269], which yielded an independent lower limit on the coupling parameter $g_{a\gamma\gamma}$ that is basic to axion physics.

Let us turn finally to the question of how much dark matter can be provided by light thermal axions of the type we have considered here. We take the rest energies given by (6.27), with $0.6 \leqslant h_0 \leqslant 0.9$ as usual. Then Eq. (6.5) yields

$$0.014 \lesssim \Omega_a \leqslant 0.053 \ . \tag{6.28}$$

This is comparable to the density of baryonic matter (Sec. 4.2), but falls well short of most expectations for the density of cold dark matter.

Our main conclusions, then, are as follows: thermal axions in the multi-eV window remain (only just) viable at the lightest end of the range of possible rest-masses given by Eq. (6.27). They may also exist with slightly higher rest-masses, up to the limit given by Eq. (6.25), but only in certain axion theories where their couplings to photons are weak. In either of these two scenarios, however, their contributions to the density of dark matter in the Universe are so feeble as to remove much of their motivation as CDM candidates. If they are to provide a significant portion of the dark matter, then axions must have rest masses in the "invisible" range where they do not contribute significantly to the light of the night sky.

Chapter 7

# The Ultraviolet Background

## 7.1 Decaying neutrinos

Experiments now indicate that neutrinos possess nonzero rest mass and make up at least part of the dark matter (Sec. 4.4). If different neutrino species have different rest masses, then heavier ones can decay into lighter ones plus a photon. These decay photons might be observable, as noticed by R. Cowsik in 1977 [270] and A. de Rujula and S.L. Glashow in 1980 [271]. The strength of the expected signal depends on the way in which neutrino masses are incorporated into the standard model of particle physics. In minimal extensions of this model, radiative neutrino decays are characterized by lifetimes on the order of $10^{29}$ yr or more [272]. This is so much longer than the age of the Universe that such neutrinos are effectively *stable*, and would not produce a detectable signal. In other theories, however, such as those involving supersymmetry, neutrinos can decay in as little as $10^{15}$ yr [273]. This is within five orders of magnitude of the age of the Universe and opens up the possibility of significant contributions to the background light.

Decay photons from neutrinos with lifetimes this short are interesting for another reason: their existence would resolve a number of longstanding astrophysical puzzles involving the ionization of hydrogen and nitrogen in the interstellar and intergalactic medium [274, 275]. As first pointed out by A.L. Melott, D.W. McKay and J.P. Ralston in 1988 [276], these would be particularly well explained by neutrinos decaying on timescales of order $\tau_\nu \sim 10^{24}$ s with rest energies $m_\nu \sim 30$ eV. This latter value fits awkwardly with current thinking on large-scale structure formation in the early Universe (Sec. 4.4). Neutrinos of this kind could help with so many other problems, however, that they have continued to draw the interest of

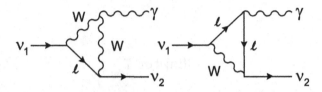

Fig. 7.1   Feynman diagrams corresponding to the decay of a massive neutrino ($\nu_1$) into a second, lighter neutrino species ($\nu_2$) together with a photon ($\gamma$). The process is mediated by charged leptons ($\ell$) and the W boson (W).

cosmologists. D.W. Sciama and his colleagues, in particular, were led on this basis to develop a detailed scenario known as the *decaying-neutrino hypothesis* [126, 277, 278]. According to this hypothesis, the rest energy and decay lifetime of the massive $\tau$-neutrino are respectively

$$m_{\nu_\tau} c^2 = 28.9 \pm 1.1 \text{ eV} \qquad \tau_\nu = (2 \pm 1) \times 10^{23} \text{ s} . \qquad (7.1)$$

The $\tau$ neutrino decays into a $\mu$ neutrino plus a photon (Fig. 7.1). Assuming that $m_{\nu_\tau} \gg m_{\nu_\mu}$, conservation of energy and momentum require this photon to have an energy $E_\gamma = \frac{1}{2} m_{\nu_\tau} c^2 = 14.4 \pm 0.5$ eV.

The concreteness of this proposal has made it eminently testable. Some of the strongest bounds come from searches for line emission near 14 eV that would be expected from concentrations of decaying dark matter in clusters of galaxies. No such signal has been seen in the direction of the galaxy cluster surrounding the quasar 3C 263 [279], or in the direction of the rich cluster Abell 665 which was observed using the Hopkins Ultraviolet Telescope (HUT) in 1991 [280]. It may be, however, that absorption plays a stronger role than expected along the lines of sight to these clusters, or that most of their dark matter is in another form [273, 281]. A potentially more robust test of the decaying-neutrino hypothesis comes from the *diffuse* background light. This has been looked at in a number of studies [282–288]. The task is a challenging one for several reasons. Decay photons of energy near 14 eV are strongly absorbed by both dust and neutral hydrogen, and the distribution of these quantities in intergalactic space is uncertain. It is also notoriously difficult, perhaps more so in this part of the spectrum than any other, to distinguish between those parts of the background which are truly extragalactic and those which are due to a complex mixture of competing foreground signals [289, 290]. We reconsider the problem here with the help of the formalism developed in Chaps. 2 and 3, adapting it to allow for absorption by gas and dust.

## 7.2 Neutrino halos

To begin with, we take as our sources of background radiation the neutrinos which have become trapped in the gravitational potential wells surrounding individual galaxies. (Not all the neutrinos will be bound in this way; and we will deal with the others separately.) The comoving number density of these galactic neutrino halos is just that of the galaxies themselves: $n_0 = 0.010\, h_0^3$ Mpc$^{-3}$ from Eq. (6.10).

The wavelength of the neutrino-decay photons at emission (like those from axion decay in Sec. 6.4) can be taken to be distributed normally about the peak wavelength corresponding to $E_\gamma$:

$$\lambda_\nu = \frac{hc}{E_\gamma} = 860 \pm 30 \text{ Å} . \tag{7.2}$$

This lies in the extreme ultraviolet (EUV) portion of the spectrum, although the redshifted tail of the observed photon spectrum will stretch across the far ultraviolet (FUV) and near ultraviolet (NUV) bands. (Universal conventions regarding the boundaries between these wavebands have yet to be established. For definiteness, we take them as follows: EUV=100–912 Å, FUV=912–2000 Å and NUV=2000–4000 Å.) The spectral energy distribution (SED) of the neutrino halos is then given by Eq. (3.19):

$$F(\lambda) = \frac{L_h}{\sqrt{2\pi}\,\sigma_\lambda} \exp\left[ -\frac{1}{2}\left( \frac{\lambda - \lambda_\gamma}{\sigma_\lambda} \right)^2 \right] , \tag{7.3}$$

where the halo luminosity $L_h$ has yet to be determined. For the standard deviation $\sigma_\lambda$ we can follow the same procedure as with axions and use the velocity dispersion in the halo, giving $\sigma_\lambda = 2\lambda_\nu v_c/c$. We parametrize this for convenience using the range of uncertainty in the value of $\lambda_\nu$, so that $\sigma_{30} \equiv \sigma_\lambda/(30 \text{ Å})$.

The halo luminosity is just the ratio of the number of decaying neutrinos ($N_\tau$) to their decay lifetime ($\tau_\nu$), multiplied by the energy of each decay photon ($E_\gamma$). Because the latter is just above the hydrogen-ionizing energy of 13.6 eV, we also need to multiply the result by an *efficiency factor* ($\epsilon$) between zero and one, to reflect the fact that some of the decay photons are absorbed by neutral hydrogen in their host galaxy before they can leave the halo and contribute to its luminosity. Altogether, then:

$$L_h = \frac{\epsilon N_\tau E_\gamma}{\tau_\nu} = \frac{\epsilon M_h c^2}{2\tau_\nu} . \tag{7.4}$$

Here we have expressed $N_\tau$ as the number of neutrinos with rest mass $m_{\nu_\tau} = 2E_\gamma/c^2$ per halo mass $M_h$.

To calculate the mass of the halo, let us follow reasoning similar to that adopted for axion halos in Sec. 6.2 and assume that the ratio of baryonic to total mass in the halo is comparable to the ratio of baryonic to total matter density in the Universe, $(M_{\text{tot}} - M_h)/M_{\text{tot}} = M_{\text{bar}}/M_{\text{tot}} = \Omega_{\text{bar}}/(\Omega_{\text{bar}} + \Omega_\nu)$. Here we have made also the economical assumption that there are no *other* contributions to the matter density, apart from those of baryons and massive neutrinos. It follows that

$$M_h = M_{\text{tot}} \left( 1 + \frac{\Omega_{\text{bar}}}{\Omega_\nu} \right)^{-1} . \tag{7.5}$$

We take $M_{\text{tot}} = (2 \pm 1) \times 10^{12} M_\odot$ following Eq. (6.13). For $\Omega_{\text{bar}}$ we use the value $(0.016 \pm 0.005) h_0^{-2}$ quoted in Sec. 4.2. And to calculate $\Omega_\nu$ we put the neutrino rest mass $m_{\nu_\tau}$ into Eq. (4.5), giving

$$\Omega_\nu = (0.31 \pm 0.01) h_0^{-2} . \tag{7.6}$$

Inserting these values of $M_{\text{tot}}$, $\Omega_{\text{bar}}$ and $\Omega_\nu$ into (7.5), we obtain

$$M_h = (0.95 \pm 0.01) M_{\text{tot}} = (1.9 \pm 0.9) \times 10^{12} M_\odot . \tag{7.7}$$

The uncertainty $h_0$ in Hubble's constant scales out of this result. Eq. (7.7) implies a baryonic mass $M_{\text{bar}} = M_{\text{tot}} - M_h \approx 1 \times 10^{11} M_\odot$, in good agreement with the observed sum of contributions from disk, bulge and halo stars, plus the matter making up the interstellar medium in our own Galaxy.

The neutrino density (7.6), when combined with that of baryons, leads to a total present-day matter density of

$$\Omega_{m,0} = \Omega_{\text{bar}} + \Omega_\nu = (0.32 \pm 0.01) h_0^{-2} . \tag{7.8}$$

As pointed out by Sciama [126], massive neutrinos are thus consistent with a critical-density Einstein-de Sitter Universe ($\Omega_{m,0} = 1$) if

$$h_0 = 0.57 \pm 0.01 . \tag{7.9}$$

This is below the range of values which many workers now consider observationally viable for Hubble's "constant" (Sec. 4.2). But it is a striking fact that the same neutrino rest mass which resolves several unrelated astrophysical problems also implies a reasonable expansion rate in the simplest cosmological model. In the interests of testing the decaying-neutrino hypothesis in a self-consistent way, we will follow Sciama and adopt the range of values (7.9) for the current chapter only.

## 7.3 Halo luminosity

To evaluate the halo luminosity (7.4), it remains to find the fraction $\epsilon$ of decay photons which escape from the halo. This problem is simplified by recognizing that the photo-ionization cross-section and distribution of neutral hydrogen in the Galaxy are such that effectively all of the decay photons striking the disk are absorbed. The probability of absorption for a single decay photon is then proportional to the solid angle subtended by the Galactic disk, as seen from the point where the photon is released.

We follow Sciama and P. Salucci [291] in modelling the distribution of $\tau$ neutrinos (and their decay photons) in the halo with a flattened ellipsoidal profile. This has

$$\rho_\nu(r,z) = 4n_\odot m_{\nu_\tau} \mathcal{N}_\nu(r,\theta)$$

$$\mathcal{N}_\nu(r,\theta) \equiv \left[ 1 + \sqrt{(r/r_\odot)^2 \sin^2\theta + (r/h)^2 \cos^2\theta} \right]^{-2} . \qquad (7.10)$$

Here $r$ and $\theta$ are spherical coordinates, $n_\odot = 5 \times 10^7$ cm$^{-3}$ is the local neutrino number density, $r_\odot = 8$ kpc is the distance of the Sun from the Galactic center, and $h = 3$ kpc is the scale height of the halo. Although this function has essentially been constructed to account for the ionization structure of the Milky Way, it agrees reasonably well with dark-matter halo distributions which have been derived on dynamical grounds [292].

Defining $x \equiv r/r_\odot$, we can use (7.10) to express the mass $M_h$ of the halo in terms of a *halo radius* $(r_h)$, thus:

$$M_h(r_h) = M_\nu \int_{\theta=0}^{\pi/2} \int_{x=0}^{x_{\max}(r_h,\theta)} \mathcal{N}_\nu(x,\theta) \, x^2 \sin\theta \, dx \, d\theta , \qquad (7.11)$$

where

$$x_{\max}(r_h,\theta) = (r_h/r_\odot)/\sqrt{\sin^2\theta + (r_\odot/h)^2 \cos^2\theta}$$

$$M_\nu \equiv 16\pi \, n_\odot m_{\nu_\tau} r_\odot^3 = 9.8 \times 10^{11} M_\odot .$$

Outside $x > x_{\max}$, we assume that the halo density drops off exponentially and can be ignored. Using (7.11) it can be shown that halos whose masses $M_h$ are given by (7.7) have scale radii $r_h = (70 \pm 25)$ kpc. This is consistent with evidence from the motion of Galactic satellites [110].

We now consider the position of a decay photon released at cylindrical coordinates $(y_\nu, z_\nu)$ inside the halo, Fig. 7.2(a). It may be seen that the disk of the Galaxy presents an approximately elliptical figure with maximum

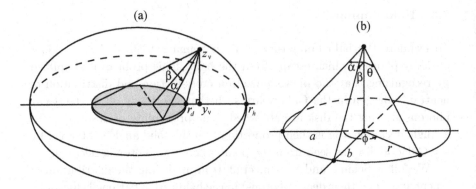

Fig. 7.2   Panel (a): absorption of decay photons inside the halo. For neutrinos at $(y_\nu, z_\nu)$, the probability that decay photons will be absorbed inside the halo (radius $r_h$) is essentially the same as the probability that they will strike the Galactic disk (radius $r_d$, shaded). Panel (b): an ellipse of semi-major axis $a$, semi-minor axis $b$ and radial arm $r(\phi)$ subtends angles of $\alpha$, $\beta$ and $\theta(\phi)$ respectively.

angular width $2\alpha$ and angular height $2\beta$, where

$$\alpha = \tan^{-1}\sqrt{(r_d^2 - d^2)/[(y_\nu - d)^2 + z_\nu^2]}$$

$$\beta = \frac{1}{2}\left[\tan^{-1}\left(\frac{y_\nu + r_d}{z_\nu}\right) - \tan^{-1}\left(\frac{y_\nu - r_d}{z_\nu}\right)\right] . \qquad (7.12)$$

Here $d = [(y_\nu^2 + z_\nu^2 + r_d^2) - \sqrt{(y_\nu^2 + z_\nu^2 + r_d^2)^2 - 4y_\nu^2 r_d^2}]/2y_\nu$, $y_\nu = r\sin\theta$, $z_\nu = r\cos\theta$ and $r_d$ is the disk radius. In spherical coordinates centered on the photon, the solid angle subtended by an ellipse is

$$\Omega_e = \int_0^{2\pi} d\phi \int_0^{\theta(\phi)} \sin\theta' \, d\theta' = \int_0^{2\pi} [1 - \cos\theta(\phi)] \, d\phi , \qquad (7.13)$$

where $\theta(\phi)$ is the angle subtended by a radial arm of the ellipse, Fig. 7.2(b). The cosine of this angle is expressed in terms of $\alpha$ and $\beta$ by

$$\cos\theta(\phi) = \left[1 + \frac{\tan^2\alpha \, \tan^2\beta}{\tan^2\alpha \, \sin^2\phi + \tan^2\beta \, \cos^2\phi}\right]^{-1/2} . \qquad (7.14)$$

The probability that a photon released at $(x, \theta)$ will escape from the halo is $\mathcal{P}_e = 1 - \Omega_e(\alpha, \beta)/4\pi$. For a given halo size $r_h$, $\epsilon$ can be found by taking the average of $\mathcal{P}_e$ over all locations $(x, \theta)$ in the halo, weighted by neutrino number density $\mathcal{N}_\nu$. Choosing $r_d = 36$ kpc as an effective disk radius [39], we obtain:

$$\epsilon = \begin{cases} 0.63 & (r_h = 45 \text{ kpc}) \\ 0.77 & (r_h = 70 \text{ kpc}) \\ 0.84 & (r_h = 95 \text{ kpc}) \end{cases} . \qquad (7.15)$$

As expected, the escape fraction of decay photons goes up as the scale size of the halo increases relative to that of the disk. For $r_h \gg r_d$ we have $\epsilon \to 1$, while a small halo with $r_h \lesssim r_d$ leads to $\epsilon \approx 0.5$.

With the decay lifetime $\tau_\nu$, halo mass $M_h$ and efficiency factor $\epsilon$ all known, Eq. (7.4) gives for the luminosity of the halo:

$$L_h = (6.5 \times 10^{42} \text{ erg s}^{-1}) f_h f_\tau^{-1} . \tag{7.16}$$

Here we have introduced two dimensionless constants $f_h$ and $f_\tau$ in order to parametrize the uncertainties in $\epsilon M_h$ and $\tau_\nu$. For the ranges of values given above, these take the values $f_h = 1.0 \pm 0.6$ and $f_\tau = 1.0 \pm 0.5$ respectively. Setting $f_h = f_\tau = 1$ gives a halo luminosity of about $2 \times 10^9 L_\odot$, or less than 5% of the optical luminosity of the Milky Way ($L_0 = 2 \times 10^{10} h_0^{-2} L_\odot$), with $h_0$ as specified by Eq. (7.9).

The combined bolometric intensity of decay photons from all the bound neutrinos out to a redshift $z_f$ is given by (2.14) as usual:

$$Q_{\text{bound}} = Q_h \int_0^{z_f} \frac{dz}{(1+z)^2 \, \tilde{H}(z)} , \tag{7.17}$$

where

$$Q_h \equiv \frac{c n_0 L_h}{H_0} = (2.0 \times 10^{-5} \text{ erg s}^{-1} \text{ cm}^{-2}) h_0^2 f_h f_\tau^{-1} .$$

The $h_0$-dependence in this quantity comes from the fact that we have so far considered only neutrinos in galaxy halos, whose number density $n_0$ goes as $h_0^3$. Since we follow Sciama in adopting the Einstein-de Sitter (EdS) cosmology in this section, the Hubble expansion rate (2.31) is

$$\tilde{H}(z) = (1+z)^{3/2} . \tag{7.18}$$

Putting this into (7.17), we find

$$Q_{\text{bound}} \approx \tfrac{2}{5} Q_h = (8.2 \times 10^{-6} \text{ erg s}^{-1} \text{ cm}^{-2}) h_0^2 f_h f_\tau^{-1} . \tag{7.19}$$

Despite their mass and size, dark-matter halos in the decaying-neutrino hypothesis are not very bright. Their combined intensity is about 1% of that of the EBL due to galaxies, $Q_* \approx 3 \times 10^{-4} \text{ erg s}^{-1} \text{ cm}^{-2}$. This is primarily due to the long decay lifetime of the neutrinos, five orders of magnitude longer than the age of the galaxies.

## 7.4   Free-streaming neutrinos

The cosmological density of decaying $\tau$ neutrinos in dark-matter halos is small: $\Omega_{\nu,\text{bound}} = n_0 M_h / \rho_{crit} = (0.068 \pm 0.032) h_0$. With $h_0$ as given by (7.9), this amounts to less than 6% of the total neutrino density, Eq. (7.6). Therefore, as expected for hot dark-matter particles, the bulk of the EBL contributions in the decaying-neutrino scenario come from particles which are distributed over larger scales. We will refer to these collectively as free-streaming neutrinos. Their cosmological density is found using (7.6) as $\Omega_{\nu,\text{free}} = \Omega_\nu - \Omega_{\nu,\text{bound}} = 0.30 h_0^{-2} f_f$, where the dimensionless constant $f_f = 1.00 \pm 0.05$ parametrizes the uncertainties in this quantity.

To identify sources of radiation in this section, we follow the same procedure as with vacuum energy (Sec. 5.5) and divide the Universe into regions of comoving volume $V_0 = n_0^{-1}$. The mass of each region is

$$M_f = \Omega_{\nu,\text{free}} \, \rho_{crit,0} V_0 = \Omega_{\nu,\text{free}} \, \rho_{crit,0} / n_0 . \qquad (7.20)$$

The luminosity of these sources has the same form as Eq. (7.4), except that we put $M_h \to M_f$ and drop the efficiency factor $\epsilon$ since the density of intergalactic hydrogen is too low to absorb a significant fraction of the decay photons within each region. Thus,

$$L_f = \frac{\Omega_{\nu,\text{free}} \, \rho_{crit,0} \, c^2}{2 n_0 \tau_\nu} . \qquad (7.21)$$

With the above values for $\Omega_{\nu,\text{free}}$ and $\tau_\nu$, and with $\rho_{crit,0}$ and $n_0$ given by (2.22) and (6.10) respectively, Eq. (7.21) implies a comoving luminosity density due to free-streaming neutrinos of

$$\mathcal{L}_f = n_0 L_f = (1.2 \times 10^{-32} \text{ erg s}^{-1} \text{ cm}^{-3}) f_f f_\tau^{-1} . \qquad (7.22)$$

This is $0.5 h_0^{-1}$ times the luminosity density of the Universe, as given by Eq. (2.18). To calculate the bolometric intensity of the background radiation due to free-streaming neutrinos, we replace $L_h$ with $L_f$ in (7.17), giving

$$Q_{\text{free}} = \frac{2 c n_0 L_f}{5 H_0} = (1.2 \times 10^{-4} \text{ erg s}^{-1} \text{ cm}^{-2}) h_0^{-1} f_f f_\tau^{-1} . \qquad (7.23)$$

This is of the same order of magnitude as $Q_*$, and goes as $h_0^{-1}$ rather than $h_0^2$. Taking into account the uncertainties in $h_0$, $f_h$, $f_f$ and $f_\tau$, the bolometric intensity of bound and free-streaming neutrinos together is

$$Q = Q_{\text{bound}} + Q_{\text{free}} = (0.33 \pm 0.17) Q_* . \qquad (7.24)$$

In principle, then, these particles are capable of shining as brightly as the galaxies themselves, Eq. (2.39). Most of this light is due to free-streaming neutrinos, which are more numerous than their halo-bound counterparts, and less affected by absorption.

## 7.5 Extinction by gas and dust

To obtain more quantitative constraints, we would like to determine neutrino contributions to the EBL as a function of wavelength. This is accomplished as in previous sections by putting the source luminosity ($L_h$ for the galaxy halos or $L_f$ for the free-streaming neutrinos) into the SED (7.3), and substituting the latter into Eq. (3.6). Now, however, we also wish to take into account the fact that decay photons encounter significant amounts of absorbing material as they travel through the *intergalactic medium* (IGM). The wavelength of neutrino decay photons, $\lambda_\nu = 860 \pm 30$ Å, is just shortward of the Lyman limit at 912 Å, which means that these photons are strongly absorbed by neutral hydrogen (this, of course, is one of the prime motivations of the Sciama theory). It is also very close to the waveband of peak extinction by dust. The simplest way to handle both these types of absorption is to include an opacity term $\tau(\lambda_0, z)$ inside the argument of the exponential. Then the intensity reads

$$I_\lambda(\lambda_0) = I_\nu \int_0^{z_f} (1+z)^{-9/2} \exp\left\{ -\frac{1}{2}\left[\frac{\lambda_0/(1+z) - \lambda_\nu}{\sigma_\lambda}\right]^2 \right.$$
$$\left. - \tau(\lambda_0, z) \right\} dz . \tag{7.25}$$

Here we have used (7.18) for $\tilde{H}(z)$. The prefactor $I_\nu$ is given with the help of (7.16) for bound neutrinos and (7.21) for free-streaming ones as

$$I_\nu = \frac{cn_0}{\sqrt{32\pi^3}H_0\sigma_\lambda} \times \begin{cases} L_h \\ L_f \end{cases} \tag{7.26}$$
$$= \begin{cases} (940 \text{ CUs})\, h_0^2\, f_h\, f_\tau^{-1}\sigma_{30}^{-1}(\lambda_0/\lambda_\nu) & \text{(bound)} \\ (5280 \text{ CUs})\, h_0^{-1}\, f_f\, f_\tau^{-1}\sigma_{30}^{-1}(\lambda_0/\lambda_\nu) & \text{(free)} \end{cases} .$$

The *optical depth* $\tau(\lambda_0, z)$ can be broken into separate terms corresponding to hydrogen gas and dust along the line of sight:

$$\tau(\lambda_0, z) = \tau_{\text{gas}}(\lambda_0, z) + \tau_{\text{dust}}(\lambda_0, z) . \tag{7.27}$$

Our best information about both of these quantities comes from observations of quasars at high redshifts. The fact that these are visible at all already places a limit on the degree of attenuation in the IGM.

We begin with the gas component. L. Zuo and E.S. Phinney [293] have developed a formalism to describe the absorption due to randomly distributed clouds such as quasar absorption-line systems, and normalized this to the number of Lyman-limit systems at $z = 3$. We use their model 1, which gives the highest absorption for $\lambda_0 \lesssim 2000$ Å, making it conservative for our purposes. Assuming an EdS cosmology, the optical depth at $\lambda_0$ due to neutral hydrogen out to a redshift $z$ is given by

$$\tau_{\text{gas}}(\lambda_0, z) = \begin{cases} \tau_{\text{ZP}} \left(\dfrac{\lambda_0}{\lambda_{\text{L}}}\right)^{3/2} \ln(1+z) & (\lambda_0 \leqslant \lambda_{\text{L}}) \\[3mm] \tau_{\text{ZP}} \left(\dfrac{\lambda_0}{\lambda_{\text{L}}}\right)^{3/2} \ln\left(\dfrac{1+z}{\lambda_0/\lambda_{\text{L}}}\right) & [\lambda_{\text{L}} < \lambda_0 < \lambda_{\text{L}}(1+z)] \\[3mm] 0 & [\lambda_0 \geqslant \lambda_{\text{L}}(1+z)] \end{cases} \quad (7.28)$$

where $\lambda_{\text{L}} = 912$ Å and $\tau_{\text{ZP}} = 2.0$.

Dust is a more complicated and potentially more important issue, and we pause to discuss this critically before proceeding. The simplest possibility, and the one which should be most effective in obscuring a diffuse signal like that considered here, would be for dust to be spread uniformly through intergalactic space. A quantitative estimate of opacity due to a uniform dusty intergalactic medium has in fact been suggested [294], but is regarded as an extreme upper limit because it would lead to excessive reddening of quasar spectra [295]. Subsequent discussions have tended to treat intergalactic dust as clumpy [296], with significant debate about the extent to which such clumps would redden and/or hide background quasars and explain a possible "turnoff" in quasar population at around $z \sim 3$ [297–300]. Most of these models assume an EdS cosmology. The effects of dust extinction could be enhanced if $\Omega_{m,0} < 1$ and/or $\Omega_{\Lambda,0} > 0$ [299], but we ignore this possibility here because we are assuming a simple model in which neutrinos make up all of the critical density.

We will use a formalism due to S.M. Fall and Y.C. Pei [301], in which dust is associated with damped Lyman-$\alpha$ absorbers, whose numbers and density profiles are sufficient to obscure a portion of the light reaching us from $z \sim 3$, but not to account fully for the "turnoff" in quasar population. Obscuration is calculated from the column density of hydrogen in these systems, together with estimates of the dust-to-gas ratio, and is normalized to the observed quasar luminosity function. The resulting mean optical depth at $\lambda_0$ out to redshift $z$ is

$$\tau_{\text{dust}}(\lambda_0, z) = \int_0^z \frac{\tau_{\text{FP}}(z')(1+z')}{(1+\Omega_{m,0}z')^{1/2}} \, \xi\left(\frac{\lambda_0}{1+z'}\right) dz'. \quad (7.29)$$

Here $\xi(\lambda)$ is the *extinction* of light by dust at wavelength $\lambda$ relative to that in the B-band (4400 Å). If $\tau_{\text{FP}}(z) = $ constant and $\xi(\lambda) \propto \lambda^{-1}$, then $\tau_{\text{dust}}$ is proportional to $\lambda_0^{-1}[(1+z)^3 - 1]$ or $\lambda_0^{-1}[(1+z)^{2.5} - 1]$, depending on cosmology [294, 296]. In the more general treatment of Fall and Pei [301], $\tau_{\rm FP}(z)$ is parametrized as a function of redshift, so that

$$\tau_{\rm FP}(z) = \tau_{\rm FP}(0)\,(1+z)^\delta, \qquad (7.30)$$

where $\tau_{\rm FP}(0)$ and $\delta$ are adjustable parameters. Assuming an EdS cosmology ($\Omega_{m,0} = 1$), the observational data are consistent with lower limits of $\tau_*(0) = 0.005$, $\delta = 0.275$ (model A); best-fit values of $\tau_*(0) = 0.016$, $\delta = 1.240$ (model B); or upper limits of $\tau_*(0) = 0.050$, $\delta = 2.063$ (model C). We will use all three models in what follows.

The shape of the extinction curve $\xi(\lambda)$ in the 300–2000 Å range can be computed using numerical Mie scattering routines in conjunction with various dust populations. Many people have constructed dust-grain models that reproduce the average extinction curve for the diffuse interstellar medium (DISM) at $\lambda > 912$ Å [302], but there have been fewer studies at shorter wavelengths. Calculations by B.T. Draine and H.M. Lee [303] have been extended into this regime by P.G. Martin and F. Rouleau [304], who assume: (1) two populations of homogeneous spherical dust grains composed of graphite and silicates respectively; (2) a power-law size distribution of the form $a^{-3.5}$ where $a$ is the grain radius; (3) a range of grain radii from 50–2500 Å; and (4) solar abundances of carbon and silicon [305].

The last of these assumptions has come into increasing question as new evidence suggests that heavy elements are far less abundant in the DISM than they are in the Sun [306]. We therefore use new dust-extinction curves based on the revised abundances [288]. In the interests of obtaining conservative bounds on the decaying-neutrino hypothesis, we consider four different grain populations, looking in particular for those that provide optimal extinction efficiency in the FUV while still reproducing the average DISM curve in the optical and NUV bands (Fig. 7.3). Our population 1 grain model (Fig. 7.3, dash-dotted line) assumes the standard grain model employed by other workers, but uses the new, lower abundance numbers together with dielectric functions due to Draine [307]. The shape of the extinction curve provides a reasonable fit to observation at longer wavelengths (reproducing for example the absorption bump at 2175 Å); but its magnitude is too low, confirming the inadequacies of the old dust model. Extinction in the vicinity of the neutrino-decay line at 860 Å is also weak, so that this model is able to "hide" very little of the light from decaying neutrinos. Insofar as it almost certainly underestimates the true extent of extinction by dust, this grain model provides a good *lower limit* on absorption in the context of the decaying-neutrino hypothesis.

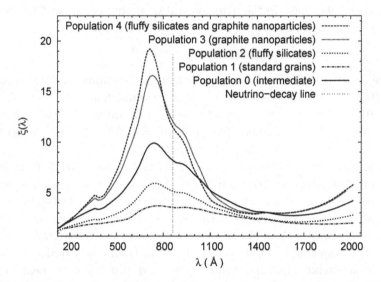

Fig. 7.3   The FUV extinction (relative to that in the B-band) produced by five different dust-grain populations. Standard grains (population 1) produce the least extinction, while PAH-like carbon nanoparticles (population 3) produce the most extinction near the neutrino-decay line at 860 Å (vertical line).

The silicate component of our population 2 grain model (Fig. 7.3, short-dashed line) is modified along the lines of the "fluffy silicate" model which has been suggested as a resolution of the heavy-element abundance problem in the DISM [308]. We replace the standard silicates of population 1 by silicate grains with a 45% void fraction, assuming a silicon abundance of $32.5 \times 10^{-6}/\mathrm{H}$ [288]. We also decrease the size of the graphite grains ($a = 50 - 250$ Å) and reduce the carbon depletion to 60%, thereby giving a better match to the DISM curve. This mixture agrees more closely with the interstellar data at optical wavelengths, and also shows significantly more FUV extinction than population 1.

For population 3 (Fig. 7.3, dotted line), we retain the standard silicates of population 1 but modify the graphite component as an approximation to the polycyclic aromatic hydrocarbon (PAH) nanostructures which have been proposed as carriers of the 2175 Å absorption bump [309]. PAH nanostructures consist of stacks of molecules such as coronene ($C_{24}H_{12}$), circumcoronene ($C_{54}H_{18}$) and larger species, in various states of edge hydrogenation. They have been linked to the 3.4 $\mu$m absorption feature in the DISM [310] as well as the extended red emission in nebular environments [311]. With sizes in the range $7 - 30$ Å, these structures are much smaller than the canonical graphite grains. Their dielectric functions, however, go over to that of graphite in the high-frequency limit [309]. So as an approximation to these particles, we use spherical graphite grains with extremely small radii ($3 - 150$ Å). This greatly increases extinction near the neutrino-decay peak.

Our population 4 grain model (Fig. 7.3, long-dashed line) combines both features of populations 2 and 3. It has the same fluffy silicate component as population 2, and the same graphite component as population 3. The results are not too different from those obtained with population 3, because extinction in the FUV waveband is dominated by small-particle contributions, so that silicates (whatever their void fraction) are of secondary importance. Neither the population 3 nor the population 4 grains fit the average DISM curve as well as those of population 2, because the Mie scattering formalism cannot accurately reproduce the behavior of nanoparticles near the 2175 Å resonance. However, the high levels of FUV extinction in these models—especially model 3 near 860 Å–suit them well for our purpose, which is to set the most conservative possible limits on the decaying-neutrino hypothesis.

## 7.6 Neutrinos and the background light

We are now ready to specify the total optical depth (7.27) and hence to evaluate the intensity integral (7.25). We will use three combinations of the dust models just described, with a view to establishing lower and upper bounds on the EBL intensity predicted by the theory. A *minimum-absorption* model is obtained by combining Fall and Pei's model A with the extinction curve of the population 1 (standard) dust grains. At the other end of the spectrum, model C of Fall and Pei together with the population 3 (nanoparticle) grains provides the most conservative *maximum-absorption* model (for $\lambda_0 \gtrsim 800$ Å). Finally, as an intermediate model, we combine model B of Fall and Pei with the extinction curve labelled as population 0 in Fig. 7.3.

The resulting predictions for the spectral intensity of the FUV background due to decaying neutrinos are plotted in Fig. 7.4 (light lines) and compared with observational limits (heavy lines and points). The curves in the bottom half of this figure refer to EBL contributions from bound neutrinos only, while those in the top half correspond to contributions from both bound and free-streaming neutrinos.

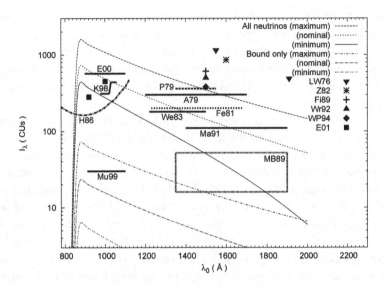

Fig. 7.4   The spectral intensity $I_\lambda$ of background radiation from decaying neutrinos as a function of observed wavelength $\lambda_0$ (light curves), plotted together with observational upper limits on EBL intensity in the far ultraviolet (points and heavy curves). The bottom four theoretical curves refer to bound neutrinos only, while the top four refer to bound and free-streaming neutrinos together. The minimum predicted signals consistent with the theory are combined with the highest possible extinction in the intergalactic medium, and vice versa.

We begin our discussion with the bound neutrinos. The key results are the three widely-spaced curves in the lower half of the figure, with peak intensities of about 6, 20 and 80 CUs at $\lambda_0 \approx 900$ Å. These are obtained by letting $h_0$ and $f_h$ take their minimum, nominal and maximum values respectively in (7.25), with the reverse order applying to $f_\tau$. Simultaneously we have adopted the maximum, intermediate and minimum-absorption models for intergalactic dust, as described above. Thus the highest-intensity model is paired with the lowest possible dust extinction, and vice versa. These curves should be seen as extreme upper and lower bounds on the theoretical intensity of EBL contributions from decaying neutrinos in galaxy halos.

These theoretical results are best compared with an experimental measurement by C. Martin and S. Bowyer in 1989 [312], labelled "MB89" in Fig. 7.4. These authors used data from a rocket-borne imaging camera to search for small-scale fluctuations in the FUV EBL, and deduced from this that the combined light of external galaxies (and their associated halos) reaches the Milky Way with an intensity of 16–52 CUs over 1350–1900 Å. There is now some doubt as to whether this was really an extragalactic signal, and indeed whether it is feasible to detect such a signal at all, given the brightness and fluctuations of the Galactic foreground in this waveband [313]. Viable or not, however, it is of interest to see what a detection of this order would mean for the decaying-neutrino hypothesis. Fig. 7.4 shows that it would constrain the theory only weakly. The expected signal in this waveband lies below 20 CUs in even the most optimistic scenario, where signal strength is highest and absorption is weakest. In the nominal "best-fit" scenario this drops to less than 7 CUs. As noted already (Sec. 7.3), the low intensity of the background light from decaying neutrinos is due to their long decay lifetimes. In order to place significant constraints on the theory, one needs the stronger signal which comes from free-streaming, as well as bound neutrinos. This in turn requires limits on the intensity of the total background rather than that associated with fluctuations.

The curves in the upper half of Fig. 7.4 (with peak intensities of about 300, 700 and 2000 CUs at $\lambda_0 \approx 900$ Å) represent the combined EBL contributions from *all* decaying neutrinos. We let $f_h$ and $f_f$ take their minimum, nominal and maximum values respectively in (7.25), with the reverse order applying to $f_\tau$ as well as $h_0$ (the latter change being due to the fact that the dominant free-streaming contribution goes as $h_0^{-1}$ rather than $h_0^2$). Simultaneously we adopt the maximum, intermediate and minimum-absorption models for intergalactic dust, as above. Intensity is reduced very significantly in the maximum-absorption case (solid line): by 11% at 900 Å, 53% at 1400 Å and 86% at 1900 Å. The bulk of this reduction is due to dust, especially at longer wavelengths where most of the light originates at high redshifts. Comparable reduction factors in the intermediate-absorption case (short-dashed line) are 9% at 900 Å, 28% at 1400 Å and 45% at 1900 Å. In the minimum-absorption case (long-dashed line), most of the extinction is due to gas rather than dust at shorter wavelengths, and intensity is reduced by a total of 9% at 900 Å, 21% at 1400 Å and 31% at 1900 Å.

The most conservative constraints on the theory are obtained by comparing the lowest predicted intensities (solid line) with observational upper limits on total EBL intensity in the FUV band (Fig. 7.4). A word is in order about these limits, which can be usefully divided into two groups: those above and below the Lyman $\alpha$-line at 1216 Å. At the longest wavelengths we include two datapoints from an analysis of OAO-2 satellite data ([57], labelled "LW76" in Fig. 7.4). Close to them is an upper limit from the Russian Prognoz satellite ([314]; "Z82"). Considerably stronger broad-band limits have come from rocket experiments by F. Paresce et al. ([315]; "P79"), R.C. Anderson et al. ([316]; "A79") and P.D. Feldman et al. ([317]; "Fe81"), as well as an analysis of data from the Solrad-11 spacecraft by C.S. Weller ([318]; "We83").

A number of other studies have proceeded by establishing a correlation between background intensity and the column density of neutral hydrogen inside the Milky Way, and then extrapolating this out to zero column density to obtain the presumed extragalactic component. Martin et al. [319] applied this correlation method to data taken by the Berkeley UVX experiment, setting an upper limit of 110 CUs on the intensity of any unidentified EBL contributions over 1400–1900 Å ("Ma91"). The correlation method is subject to uncertainties involving the true extent of scattering by dust, as well as absorption by ionized and molecular hydrogen at high Galactic latitudes. R.C. Henry and J. Murthy [290, 320] approach these issues differently and raise the upper limit on background intensity to 400 CUs over 1216–3200 Å. A good indication of the complexity of the problem is found at 1500 Å, where data from the DE-1 satellite has been used to identify an isotropic background flux of 530 ± 80 CUs ([321]; "Fi89"). The same data were subsequently reanalyzed to give a best-fit value of 45 CUs, with a conservative upper limit of 500 CUs ([322]; "Wr92"). The first of these numbers would rule out the decaying-neutrino hypothesis, while the latter does not constrain it at all. A third treatment of the same data has led to an intermediate result of 300 ± 80 CUs ([323]; "WP94").

Limits on the FUV background shortward of Lyα have been even more controversial. Several studies have been based on data from the Voyager 2 ultraviolet spectrograph, beginning with that of J.B. Holberg [324], who obtained limits between 100 and 200 CUs over 500–1100 Å (labelled "H86" in Fig. 7.4). An analysis of the same data over 912–1100 Å by Murthy et al. [325] led to similar numbers. In a subsequent reanalysis, however, Murthy et al. [326] tightened this bound to 30 CUs over the same wave-band ("Mu99"). The statistical validity of these results has been vigorously debated [327, 328], with a second group asserting that the original data do not justify a limit smaller than 570 CUs ("E00"). Of these Voyager-based limits, the strongest ("Mu99") is incompatible with the decaying-neutrino hypothesis, while the weakest ("E00") constrains it only weakly. Two more recent experiments yield results midway between these extremes: the DUVE orbital spectrometer [329] and the EURD spectrograph aboard the Spanish MINISAT 01 [330]. Upper limits on continuum emission from the former instrument are 310 CUs over 980–1020 Å and 440 CUs over 1030–1060 Å ("K98"); while the latter has produced upper bounds of 280 CUs at 920 Å and 450 CUs at 1000 Å ("E01").

The above observational data impose strong constraints on the decaying-neutrino hypothesis, in which neutrinos make up nearly all of the dark matter. Longward of Lyα, Fig. 7.4 shows that the data span very nearly the same parameter space as the minimum and maximum-intensity predictions of the theory (solid and long-dashed lines). Most stringent are Weller's Solrad-11 result ("We83") and the correlation-method constraint of Martin *et al.* ("Ma91"). Absorption (by dust in particular) plays a critical role in reducing the strength of the signal.

Shortward of Lyα, most of the signal originates nearby, and intergalactic absorption is far less important. Ambiguity here comes rather from the spread in reported limits, which in turn reflects the formidable experimental challenges in this part of the spectrum. Nevertheless it is clear that both the Voyager-based limits of Holberg ("H86") and Murthy *et al.* ("Mu99"), as well as the new EURD measurement at 920 Å ("E01"), are incompatible with the theory. These upper bounds are violated by even the weakest predicted signal, which assumes the strongest possible extinction (solid line). The easiest way to reconcile theory with observation is to increase the neutrino decay lifetime. If we require that $I_{th} < I_{obs}$, then the abovementioned EURD measurement ("E01") implies a lower bound of $\tau_\nu > 3 \times 10^{23}$ s. This rises to $(5 \pm 3) \times 10^{23}$ s and $(26 \pm 10) \times 10^{23}$ s for the Voyager limits ("H86" and "Mu99" respectively). All these numbers lie outside the range of lifetimes required in the decaying-neutrino scenario, $\tau_\nu = (2 \pm 1) \times 10^{23}$ s. The DUVE constraint ("K98") is more forgiving, but still pushes the theory to the edge of its available parameter space. *Taken together, these data cast serious doubt on the viability of the decaying-neutrino hypothesis.* This conclusion is in accord with current thinking on the value of Hubble's constant (Sec. 4.2) and structure formation (Sec. 4.4), as well as more detailed analysis of the EURD data [331].

These limits would be weakened (by a factor of up to nearly one-third) if the value of Hubble's constant $h_0$ were allowed to exceed $0.57 \pm 0.01$, since the dominant free-streaming contributions to $I_\lambda(\lambda_0)$ go as $h_0^{-1}$. A higher expansion rate would however exacerbate problems with structure formation and the age of the Universe, the more so because the dark matter in this theory is hot. It would also mean sacrificing the critical density of neutrinos. Another possibility would be to consider lower neutrino rest masses, a scenario that does not conflict with other observational data until $m_\nu c^2 \lesssim 2$ eV [332]. This would however entail a proportional reduction in decay photon energy, which would have to drop below the Lyman or hydrogen-ionizing limit, thus removing the whole motivation for the proposed neutrinos in the first place. Similar considerations apply to neutrinos with longer decay lifetimes.

Our conclusions are as follows. Neutrinos with rest masses and decay lifetimes as specified by the decaying-neutrino scenario produce levels of ultraviolet background radiation very close to, and in some cases above, observational limits on the EBL. At wavelengths longer than 1200 Å, where intergalactic absorption is most effective, the theory is marginally compatible with observation—*if* one adopts the upper limits on dust density consistent with quasar obscuration, and *if* the dust grains are extremely small. At wavelengths in the range 900–1200 Å, no such accommodation is possible, and the theory predicts a background intensity either comparable to or significantly higher than observational limits. Thus, while we know that there are many neutrinos out there, the gloom of the ultraviolet night sky agrees with other evidence in telling us that they cannot make up a significant fraction of the dark matter.

Chapter 8

# The X-ray and Gamma-ray Backgrounds

## 8.1 Weakly interacting massive particles

Weakly interacting massive particles (WIMPs) are as-yet undiscovered particles whose rest masses far exceed those of baryons, but whose interaction strengths are comparable to those of neutrinos. The most widely-discussed examples arise in the context of supersymmetry (SUSY). This is a theoretical concept, motivated in broad terms by the wish to unify gravity with the stronger interactions of particle physics. For the latter, standard calculations based on quantum field theory lead to extremely large values of vacuum (or zero-point) fields, incompatible with the small values inferred from the global cosmological constant (Chap. 5). With SUSY, these particle fields are largely cancelled, by partnering each integral-spin boson with a hypothetical half-integer-spin fermion, and vice versa. (This more than doubles the number of fundamental parameters in the theory; see [333] for a review.) These superpartners were recognized as potential dark-matter candidates in the early 1980s by N. Cabibbo *et al.* [334], H. Pagels and J.R. Primack [335], S. Weinberg [336] and others [337–341], with the generic term "WIMP" being coined in 1985 [342].

There is, as yet, no firm experimental evidence for SUSY WIMPs. This means that their rest energies, if they exist, lie beyond the range currently probed by accelerators (and in particular beyond the rest energies of their standard-model counterparts). Supersymmetry is, therefore, not an exact symmetry of nature. The masses of the superpartners, like that of the axion (Chap. 6), must have been generated by a symmetry breaking in the early Universe. Subsequently, as the temperature of the expanding fireball dropped below their rest energies, heavier species would have dropped out of equilibrium and begun to disappear by pair annihilation, leaving progressively lighter ones behind. Eventually, only one massive superpartner would have remained: the *lightest supersymmetric particle* (LSP). It is this particle which plays the role of the WIMP in SUSY theories. Calculations using the Boltzmann equation show that the relic LSP density today lies within one or two orders of magnitude of the required density of CDM across much of the SUSY parameter space [343]. In this respect, SUSY WIMPs are more natural DM candidates than axions (Chap. 6), whose cosmological density ranges *a priori* over many orders of magnitude.

SUSY WIMPs contribute to the cosmic background radiation in at least three ways. The first is by pair annihilation to photons. This process occurs even in the simplest, or minimal SUSY model (MSSM), but is very slow because it takes place via intermediate loops of charged particles such as leptons and quarks and their antiparticles. The underlying reason for the stability of the LSP in the MSSM is an additional new symmetry of nature, known as R-parity, which is necessary (among other things) to protect the proton from decaying via intermediate SUSY states. The other two types of background contributions occur in *non*-minimal SUSY theories, in which R-parity is not conserved (and in which the proton can decay). In these theories, LSPs can decay into photons directly via loop diagrams, and also indirectly via tree-level decays to secondary particles which then scatter off pre-existing background photons to produce a signal.

The first step in assessing the importance of each of these processes is to identify the LSP. Early candidates included photinos ($\tilde{\gamma}$) and gravitinos ($\tilde{g}$), the fermionic superpartners of photons [334] and gravitons [335], as well as sneutrinos ($\tilde{\nu}$) and selectrons ($\tilde{e}$), the bosonic counterparts of neutrinos [339] and electrons [340]. (SUSY particles are denoted by tildes and take the same names as their standard-model partners, with a prefix "s" for superpartners of fermions and a suffix "ino" for those of bosons.) However, as demonstrated by J. Ellis and others in 1984 [341], most of these possibilities encounter problems. The LSP is in fact most likely to be a *neutralino* ($\tilde{\chi}$), a linear superposition of the photino ($\tilde{\gamma}$), the zino ($\tilde{Z}$) and two neutral higgsinos ($\tilde{h}_1^0$ and $\tilde{h}_2^0$). (These are the SUSY spin-$\frac{1}{2}$ counterparts of the photon, $Z^0$ and Higgs bosons respectively.) There are four neutralinos, each a mass eigenstate made up of (in general) different amounts of photino, zino, etc., although in special cases a neutralino could be "mostly photino," say, or "pure zino." The LSP is by definition the lightest such eigenstate. Accelerator searches place a lower limit on its rest energy which currently stands at $m_{\tilde{\chi}}c^2 > 46$ GeV [344].

In minimal SUSY, the density of neutralinos drops only by way of the (slow) pair-annihilation process, so that these particles could "overclose" the Universe if their rest energy is too high. This speeds up the expansion rate of the Universe, which is proportional to the square root of the total matter density from Eq. (2.20). Lower bounds on the age of the Universe thus impose an upper bound on the neutralino rest energy which has been set at $m_{\tilde{\chi}}c^2 \lesssim 3200$ GeV [345]. Detailed exploration of the parameter space of minimal SUSY theory tightens this upper limit in most cases to $m_{\tilde{\chi}}c^2 \lesssim 600$ GeV [346]. Recent work has focused on a slimmed-down version of the MSSM known as the *constrained minimal SUSY model* (CMSSM), in which all existing experimental bounds and cosmological requirements are comfortably met by neutralinos with rest energies in the range 90 GeV$\lesssim m_{\tilde{\chi}}c^2 \lesssim 400$ GeV [347].

Even in its constrained minimal version, SUSY physics contains at least five adjustable input parameters, making the neutralino a harder proposition to test than the axion or the massive neutrino. Fortunately, there are several other ways (besides accelerator searches) to look for these particles. Because their rest energies are above the temperature at which they decoupled from the primordial fireball, WIMPs have non-relativistic velocities and are found predominantly in gravitational potential wells like those of our own Galaxy. They will occasionally scatter against target nuclei in terrestrial detectors as the Earth follows the Sun around the Milky Way. Annual variations in this signal resulting from the Earth's orbital motion through the Galactic dark-matter halo can be used to isolate a WIMP signal. Just such a signal was reported by the DAMA experiment in 2000 using detectors in Italy, with an implied WIMP rest energy of $m_{\tilde{\chi}}c^2 = 52^{+10}_{-8}$ GeV [348]. However, subsequent experiments in France (EDELWEISS [349]) and the U.S.A. (CDMS [350, 351]) have not been able to reproduce this result. New detectors such as ZEPLIN in England [352] and IGEX in Spain [353] are coming online to help with the search.

A second, indirect search strategy is to look for annihilation byproducts from neutralinos which have collected inside massive bodies. Most attention has been directed at the possibility of detecting antiprotons from the Galactic halo [354] or neutrinos from the Sun [355] or Earth [356]. The heat generated in the cores of gas giants like Jupiter or Uranus has also been considered as a potential annihilation signature [357]. The main challenge in each case lies in separating the signal from the background noise. In the case of the Earth, one can look for neutrino-induced muons which are distinguishable from the atmospheric background by the fact that they are travelling straight up. The AMANDA experiment, whose detectors are buried deep in the Antarctic ice, has recently reported upper limits on the density of terrestrial WIMPs based on this principle [358].

## 8.2 Pair annihilation

Pair annihilation into *photons* provides a complementary indirect search technique. For the WIMP rest energies considered here ($50 \text{ GeV} \lesssim m_{\tilde{\chi}}c^2 \lesssim 1000 \text{ GeV}$), the photons so produced lie in the gamma-ray portion of the spectrum. Beginning with D.W. Sciama [340] and J. Silk and M. Srednicki [354], many workers have studied the possibility of gamma-rays from SUSY WIMP annihilations in the halo of the Milky Way. Prognoses for detection have ranged from very optimistic [359] to very pessimistic [360]; converging gradually to the conclusion that neutralino-annihilation contributions would be at or somewhat below the level of the Galactic background, and possibly distinguishable from it by their spectral shape [361–363]. Recent studies have focused on possible enhancements of the signal in the presence of a high-density Galactic core [364], a flattened halo [365], a very extended singular halo [366], a massive central black hole [367], significant substructure [368–370] and adiabatic compression due to baryons [371]. A good summary is found in Ref. [372] with attention to prospects for detection by the upcoming GLAST mission.

WIMPs with rest energies at the upper end of the above-mentioned range ($1 \text{ TeV} \equiv 10^{12} \text{ eV}$) would produce a weaker signal, but this might be compensated by the larger effective area of the atmospheric Čerenkov telescopes (ACTs) used to detect them [374]. Not all authors are as sanguine [375], but observations of high-energy gamma-rays from the Galactic center by the CANGAROO [376] and VERITAS collaborations [377] provide tantalizing examples of what might be possible with this technique. The Milagro extensive air-shower array is another experiment that has recently set upper limits on the density of TeV-energy WIMPs in the vicinity of the Sun [378]. Others have carried the search farther afield, toward objects like dwarf spheroidal galaxies [379], the Large Magellanic Cloud [380] and the giant elliptical galaxy M87 in Virgo [381].

SUSY WIMP contributions to the *diffuse* extragalactic background (as opposed to localized sources) have received less attention. First to study the problem were Cabibbo *et al.* [334], who however assumed a WIMP rest energy (10–30 eV) which we now know is far too low. Like the decaying neutrino (Chap. 7), such a particle would produce a signal in the ultravi-

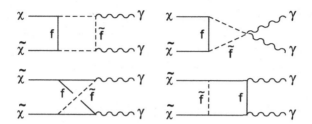

Fig. 8.1  Feynman diagrams corresponding to the annihilation of two neutralinos ($\tilde{\chi}$), producing a pair of photons ($\gamma$). The process is mediated by fermions ($f$) and their supersymmetric counterparts, the sfermions ($\tilde{f}$).

olet. It is excluded, however, by an argument known as the Lee-Weinberg bound, which effectively restricts WIMPs to rest energies above 2 GeV [338]. The background light from WIMPs in this range was first estimated by Silk and Srednicki [354]. Their conclusion, and that of most workers who have followed them [382–384], is that neutralino annihilations could be responsible for no more than a small fraction of the observed gamma-ray background. We review this argument below, reversing our usual procedure and attempting to set a reasonably conservative *upper* limit on neutralino contributions to the EBL.

Neutralino annihilations produce most of their light indirectly, via tree-level annihilations to hadrons (mostly pions) which then decay to photons, electrons, positrons and neutrinos. (The electrons and positrons add even more to the signal by inverse Compton scattering off low-energy CMB photons.) However, the energies of the photons produced in this way are broadly distributed, resulting in a continuum gamma-ray spectrum which is difficult if not impossible to distinguish from the astrophysical background [385]. We therefore concentrate on processes in which neutralino pairs annihilate *directly* into photon pairs via loop diagrams, as shown in Fig. 8.1. These give rise to a photon spectrum that is essentially *monoenergetic*, $E_\gamma \approx m_{\tilde{\chi}} c^2$. No conventional astrophysical processes produce such a narrow peak, whose detection against the diffuse extragalactic background would constitute compelling evidence for dark matter.

We again take galactic dark-matter halos as our sources of background radiation, with comoving number density $n_0$. Photon wavelengths are distributed normally about the peak wavelength in the galaxy rest frame:

$$\lambda_{\mathrm{ann}} = hc/m_{\tilde{\chi}} c^2 = (1.2 \times 10^{-6} \text{ Å}) \, m_{10}^{-1} \, , \qquad (8.1)$$

where $m_{10} \equiv m_{\tilde{\chi}} c^2/(10 \text{ GeV})$ is the neutralino rest energy in units of 10 GeV. The standard deviation $\sigma_\gamma$ can be related to the velocity dispersion of bound dark-matter particles as in previous sections, so that $\sigma_\lambda = 2(v_c/c)\lambda_{ann}$. With $v_c \sim 220$ km s$^{-1}$ and $m_{10} \sim 1$, this is of order $\sim 10^{-9}$ Å. For convenience we specify this with the dimensionless parameter $\sigma_9 \equiv \sigma_\lambda/(10^{-9}$ Å). The spectral energy distribution is then given by Eq. (3.19) as

$$F(\lambda) = \frac{L_{h,ann}}{\sqrt{2\pi}\,\sigma_\lambda} \exp\left[ -\frac{1}{2}\left(\frac{\lambda - \lambda_{ann}}{\sigma_\lambda}\right)^2 \right] . \tag{8.2}$$

The luminosity due to neutralino annihilations is proportional to the rest energy of the annihilating particles times the annihilation rate, which in turn goes as the cross-section $(\sigma v)$ times the square of the neutralino number density, $n_{\tilde\chi}^2$. The resulting expression may be written

$$L_{h,ann} = 2m_{\tilde\chi}c^2 <\sigma v>_{\gamma\gamma} \mathcal{P} , \tag{8.3}$$

where $<\sigma v>_{\gamma\gamma}$ is the photo-annihilation cross-section, the quantity $\mathcal{P} \equiv 4\pi m_{\tilde\chi}^{-2} \int \rho_{\tilde\chi}^2(r)r^2 dr$ is the radial average of $n_{\tilde\chi}^2$ over the halo and $\rho_{\tilde\chi}(r)$ is the neutralino density distribution. V.S. Berezinsky *et al.* [386] have determined $<\sigma v>_{\gamma\gamma} \approx a_{\gamma\gamma}$ for non-relativistic neutralinos as

$$a_{\gamma\gamma} = \frac{\hbar^2 c^3 \alpha^4 (m_{\tilde\chi}c^2)^2}{3^6 \pi (m_{\tilde f}c^2)^4} \left(\frac{Z_{11}}{\sin\theta_W}\right)^4 (45 + 48y + 139y^2)^2 . \tag{8.4}$$

Here $\alpha$ is the fine structure constant, $m_{\tilde f}$ is the mass of an intermediate sfermion, $y \equiv (Z_{12}/Z_{11})\tan\theta_W$, $\theta_W$ is the weak mixing angle and $Z_{ij}$ are elements of the real orthogonal matrix which diagonalizes the neutralino mass matrix. In particular, the "pure photino" case is specified by $Z_{11} = \sin\theta_W, y = 1$ and the "pure zino" by $Z_{11} = \cos\theta_W, y = -\tan^2\theta_W$. Collecting these expressions together and parametrizing the sfermion rest energy by $\tilde m_{10} \equiv m_{\tilde f}/10$ GeV, we obtain:

$$<\sigma v>_{\gamma\gamma} = (8 \times 10^{-27}\text{ cm}^3\text{s}^{-1})f_\chi m_{10}^2 \tilde m_{10}^{-4} . \tag{8.5}$$

Here $f_\chi$ (=1 for photinos, 0.4 for zinos) is a dimensionless quantity whose value parametrizes the makeup of the neutralino.

Since we attempt in this section to set an upper limit on EBL contributions from neutralino annihilations, we take $f_\chi \approx 1$ (the photino case). In the same spirit, we would like to use lower limits for the sfermion mass $\tilde m_{10}$. It is important to estimate this quantity accurately since the cross-section goes as $\tilde m_{10}^{-4}$. G.F. Giudice and K. Griest [387] have made a detailed study of photino annihilations and find a lower limit on $\tilde m_{10}$ as a function of $m_{10}$, assuming that photinos provide at least $0.025h_0^{-2}$ of the critical density.

Over the range $0.1 \leqslant m_{10} \leqslant 4$, this lower limit is empirically well fit by a function of the form $\tilde{m}_{10} \approx 4m_{10}^{0.3}$. If this holds over our broader range of masses, then we obtain an upper limit on the neutralino annihilation cross-section of $<\sigma v>_{\gamma\gamma} \lesssim (3 \times 10^{-29} \text{ cm}^3\text{s}^{-1}) m_{10}^{0.8}$.

For the halo density profile $\rho_{\tilde{\chi}}(r)$ we adopt the simple and widely-used *isothermal model* [363]:

$$\rho_{\tilde{\chi}}(r) = \rho_{\odot} \left( \frac{a^2 + r_{\odot}^2}{a^2 + r^2} \right) . \tag{8.6}$$

Here $\rho_{\odot} = 5 \times 10^{-25}$ g cm$^{-3}$ is the density of dark matter in the solar system, assuming a spherical halo [292], $r_{\odot} = 8$ kpc is the distance of the Sun from the Galactic center [388] and $a = (2 - 20)$ kpc is a core radius. To fix this latter parameter, we can integrate (8.6) over volume to obtain total halo mass $M_h(r)$ inside radius $r$:

$$M_h(r) = 4\pi\rho_{\odot}r(a^2 + r_{\odot}^2) \left[ 1 - \left(\frac{a}{r}\right) \tan^{-1} \left(\frac{r}{a}\right) \right] . \tag{8.7}$$

Observations of the motions of Galactic satellites imply that the total mass inside 50 kpc is about $5 \times 10^{11} M_{\odot}$ [110]. This in (8.7) implies $a = 9$ kpc, which we consequently adopt. The maximum extent of the halo is not well-constrained observationally, but can be specified if we take $M_h = (2 \pm 1) \times 10^{12} M_{\odot}$ as in (6.13). Eq. (8.7) then gives a halo radius $r_h = (170 \pm 80)$ kpc. The cosmological density of WIMPs in galactic dark-matter halos adds up to $\Omega_h = n_0 M_h / \rho_{\text{crit},0} = (0.07 \pm 0.04) h_0$.

If there are no other sources of CDM, then the total matter density is $\Omega_{m,0} = \Omega_h + \Omega_{\text{bar}} \approx 0.1 h_0$ and the observed flatness of the Universe (Chap. 4) implies a strongly vacuum-dominated cosmology. While we use this as a lower limit on WIMP contributions to the dark matter in subsequent sections, it is likely that CDM also exists in larger-scale regions such as galaxy clusters. To take this into account in a general way, we define a *cosmological enhancement factor* $f_c \equiv (\Omega_{m,0} - \Omega_{\text{bar}})/\Omega_h$ representing the added contributions from WIMPs outside galactic halos (or perhaps in halos which extend far enough to fill the space between galaxies). This takes the value $f_c = 1$ for the most conservative case just described, but rises to $f_c = (4 \pm 2) h_0^{-1}$ in the $\Lambda$CDM model with $\Omega_{m,0} = 0.3$, and $(14 \pm 7) h_0^{-1}$ in the EdS model with $\Omega_{m,0} = 1$.

With $\rho_{\tilde{\chi}}(r)$ known, we are in a position to calculate the quantity $\mathcal{P}$:

$$\mathcal{P} = \frac{2\pi\rho_{\odot}^2(a^2 + r_{\odot}^2)^2}{m_{\tilde{\chi}}^2 a} \left[ \tan^{-1} \left(\frac{r_h}{a}\right) - \frac{(r_h/a)}{1 + (r_h/a)^2} \right] . \tag{8.8}$$

Using the values for $\rho_\odot$, $r_\odot$ and $a$ specified above and setting $r_h = 250$ kpc to get an upper limit, we find that $\mathcal{P} \leqslant (5 \times 10^{65} \text{ cm}^{-3}) m_{10}^{-2}$. Putting this result along with the cross-section (8.5) into (8.3), we obtain:

$$L_{h,\text{ann}} \leqslant (1 \times 10^{38} \text{ erg s}^{-1}) f_\chi m_{10} \tilde{m}_{10}^{-4} . \tag{8.9}$$

Inserting Giudice and Griest's [387] lower limit on the sfermion mass $\tilde{m}_{10}$ (as empirically fit above), we find that (8.9) gives an upper limit on halo luminosity of $L_{h,\text{ann}} \leqslant (5 \times 10^{35} \text{ erg s}^{-1}) f_\chi m_{10}^{-0.2}$. Higher estimates can be found in the literature [389], but these assume a singular halo whose density drops off as only $\rho_{\tilde{\chi}}(r) \propto r^{-1.8}$ and extends out to a very large halo radius, $r_h = 4.2 h_0^{-1}$ Mpc. For a standard isothermal distribution of the form (8.6), our results confirm that halo luminosity due to neutralino annihilations alone is very low, amounting to less than $10^{-8}$ times the total bolometric luminosity of the Milky Way.

The combined bolometric intensity of neutralino annihilations between redshift $z_f$ and the present is given by substituting the comoving number density $n_0$ and luminosity $L_{h,\text{ann}}$ into Eq. (2.13). This gives

$$Q = Q_{\tilde{\chi},\text{ann}} \int_0^{z_f} \frac{dz}{(1+z)^2 \left[\Omega_{m,0}(1+z)^3 + (1 - \Omega_{m,0})\right]^{1/2}} , \tag{8.10}$$

where $Q_{\tilde{\chi},\text{ann}} = (cn_0 L_{h,\text{ann}} f_c)/H_0$ and we have assumed spatial flatness. With values for the parameters as specified above, we find

$$Q = \begin{cases} (1 \times 10^{-12} \text{ erg s}^{-1} \text{ cm}^{-2}) h_0^2 f_\chi m_{10}^{-0.2} & (\text{if } \Omega_{m,0} = 0.1 h_0) \\ (3 \times 10^{-12} \text{ erg s}^{-1} \text{ cm}^{-2}) h_0 f_\chi m_{10}^{-0.2} & (\text{if } \Omega_{m,0} = 0.3) \\ (1 \times 10^{-11} \text{ erg s}^{-1} \text{ cm}^{-2}) h_0 f_\chi m_{10}^{-0.2} & (\text{if } \Omega_{m,0} = 1) \end{cases} . \tag{8.11}$$

Here we have set $z_f = 30$ and chosen $f_c = 1$, $4 h_0^{-1}$ and $20 h_0^{-1}$ respectively. The effect of a larger cosmological enhancement factor $f_c$ is partially offset in (8.10) by the fact that a Universe with higher matter density $\Omega_{m,0}$ is younger, and hence contains less background light in general. But even the highest value of $Q$ given in (8.11) is negligible in comparison to the intensity (2.19) of the EBL due to ordinary galaxies.

The total spectral intensity of annihilating neutralinos is found by substituting the SED (8.2) into (3.6) to give

$$I_\lambda(\lambda_0) = I_{\tilde{\chi},\text{ann}} \int_0^{z_f} \frac{\exp\left\{-\frac{1}{2}\left[\dfrac{\lambda_0/(1+z) - \lambda_{\text{ann}}}{\sigma_\lambda}\right]^2\right\} dz}{(1+z)^3 \left[\Omega_{m,0}(1+z)^3 + 1 - \Omega_{m,0}\right]^{1/2}} . \tag{8.12}$$

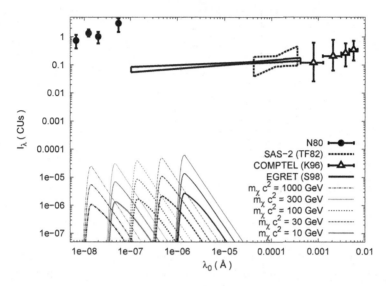

Fig. 8.2 The spectral intensity of the diffuse gamma-ray background due to neutralino annihilations (lower left), compared with observational limits from high-altitude balloon experiments (N80), the SAS-2 spacecraft and the COMPTEL and EGRET instruments. The three plotted curves for each value of $m_{\tilde{\chi}}c^2$ depend on the total density of neutralinos: galaxy halos only ($\Omega_{m,0} = 0.1h_0$; heavy lines), $\Lambda$CDM model ($\Omega_{m,0} = 0.3$; medium lines), or EdS model ($\Omega_{m,0} = 1$; light lines).

For a typical neutralino with $m_{10} \approx 10$ the annihilation spectrum peaks near $\lambda_0 \approx 10^{-7}$ Å. The dimensional prefactor reads

$$
I_{\tilde{\chi},\mathrm{ann}} = \frac{n_0 L_{h,\mathrm{ann}} f_c}{\sqrt{32\pi^3}\, h\, H_0} \left(\frac{\lambda_0}{\sigma_\lambda}\right)
$$

$$
= (0.0002 \text{ CUs})\, h_0^2\, f_\chi\, f_c\, m_{10}^{0.8}\, \sigma_9^{-1} \left(\frac{\lambda_0}{10^{-7} \text{ Å}}\right) . \tag{8.13}
$$

Here we have divided through by the photon energy $hc/\lambda_0$ to put results into continuum units or CUs as usual (Sec. 3.2). Eq. (8.12) gives the combined intensity of radiation from neutralino annihilations, emitted at various wavelengths and redshifted by various amounts, but observed at wavelength $\lambda_0$. Results are plotted in Fig. 8.2 together with observational constraints. We defer discussion of this plot (and the experimental data) to Sec. 8.6, for easier comparison with the results from other WIMP-related processes.

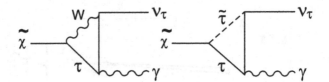

Fig. 8.3    Feynman diagrams corresponding to one-loop decays of the neutralino ($\tilde{\chi}$) into a neutrino (here the $\tau$-neutrino) and a photon ($\gamma$). The process can be mediated by the W-boson and a $\tau$-lepton, or by the $\tau$ and its supersymmetric counterpart ($\tilde{\tau}$).

## 8.3    One-loop decay

We turn next to *non*-minimal SUSY theories in which R-parity is not necessarily conserved and the LSP (in this case the neutralino) can decay [390]. There is one *direct* decay mode into photons, $\tilde{\chi} \to \nu + \gamma$ (Fig. 8.3). Because these decays occur via loop diagrams, they are again subdominant. We consider theories in which R-parity breaking is accomplished spontaneously. This means introducing an intermediate scalar sneutrino with a nonzero vacuum expectation value $v_R \equiv <\tilde{\nu}_{\tau_R}>$ [391]. Neutralino decays into photons could be detectable if $m_{\tilde{\chi}}$ and $v_R$ are large [392].

The photons produced in this way are again monochromatic, with $E_\gamma = \frac{1}{2} m_{\tilde{\chi}} c^2$. In fact the SED here is the same as (8.2) except that peak wavelength is doubled, $\lambda_{\mathrm{loop}} = 2hc/m_{\tilde{\chi}} c^2 = (2.5 \times 10^{-6} \, \text{Å}) \, m_{10}^{-1}$. The only parameter that needs to be recalculated is the halo luminosity $L_h$. For one-loop neutralino decays of lifetime $\tau_{\tilde{\chi}}$, this takes the form:

$$L_{h,\mathrm{loop}} = \frac{N_{\tilde{\chi}} b_\gamma E_\gamma}{\tau_{\tilde{\chi}}} = \frac{b_\gamma M_h c^2}{2 \tau_{\tilde{\chi}}} \; . \tag{8.14}$$

Here $N_{\tilde{\chi}} = M_h/m_{\tilde{\chi}}$ is the number of neutralinos in the halo and $b_\gamma$ is the *branching ratio*, or fraction of neutralinos that decay into photons. This is estimated by Berezinsky *et al.* [392] as

$$b_\gamma \approx 10^{-9} f_R^2 m_{10}^2 \; , \tag{8.15}$$

where the new parameter $f_R \equiv v_R/(100 \text{ GeV})$. The requirement that SUSY WIMPs not carry too much energy out of stellar cores implies that $f_R$ is of order ten or more [391]. We take $f_R > 1$ as a lower limit.

We adopt $M_h = (2\pm1) \times 10^{12} M_\odot$ as usual, with $r_h = (170\pm80)$ kpc from the discussion following (8.7). As in the previous section, we parametrize our lack of certainty about the distribution of neutralinos on larger scales with the cosmological enhancement factor $f_c$. Collecting these results together and expressing the decay lifetime in dimensionless form as $f_\tau \equiv \tau_{\tilde\chi}/(1 \text{ Gyr})$, we obtain for the halo luminosity:

$$L_{h,\text{loop}} = (6 \times 10^{40} \text{ erg s}^{-1}) m_{10}^2 f_R^2 f_\tau^{-1} . \tag{8.16}$$

With $m_{10} \sim f_R \sim f_\tau \sim 1$, Eq. (8.16) gives $L_{h,\text{loop}} \sim 2 \times 10^7 L_\odot$. This is considerably brighter than the halo luminosity due to neutralino annihilations in minimal SUSY models, but still amounts to less than $10^{-3}$ times the bolometric luminosity of the Milky Way.

Combined bolometric intensity is found as in the previous section, but with $L_{h,\text{ann}}$ in (8.10) replaced by $L_{h,\text{loop}}$ so that

$$Q = \begin{cases} (1 \times 10^{-7} \text{ erg s}^{-1} \text{ cm}^{-2}) h_0^2 m_{10}^2 f_R^2 f_\tau^{-1} & (\text{if } \Omega_{m,0} = 0.1h_0) \\ (4 \times 10^{-7} \text{ erg s}^{-1} \text{ cm}^{-2}) h_0 m_{10}^2 f_R^2 f_\tau^{-1} & (\text{if } \Omega_{m,0} = 0.3) \\ (2 \times 10^{-6} \text{ erg s}^{-1} \text{ cm}^{-2}) h_0 m_{10}^2 f_R^2 f_\tau^{-1} & (\text{if } \Omega_{m,0} = 1) \end{cases} . \tag{8.17}$$

This is again small. However, we see that massive ($m_{10} \gtrsim 10$) neutralinos which provide close to the critical density ($\Omega_{m,0} \sim 1$) and decay on timescales of order 1 Gyr or less ($f_\tau \lesssim 1$) could in principle rival the intensity of the conventional EBL.

To obtain more quantitative constraints, we turn to spectral intensity. This is given by Eq. (8.12) as before, except that the dimensional prefactor $I_{\tilde\chi,\text{ann}}$ must be replaced by

$$I_{\tilde\chi,\text{loop}} = \frac{n_0 L_{h,\text{loop}} f_c}{\sqrt{32\pi^3} h H_0} \left(\frac{\lambda_0}{\sigma_\lambda}\right)$$
$$= (30 \text{ CUs}) h_0^2 m_{10}^3 f_R^2 f_\tau^{-1} f_c \sigma_9^{-1} \left(\frac{\lambda_0}{10^{-7} \text{ Å}}\right) . \tag{8.18}$$

Results are plotted in Fig. 8.4 for neutralino rest energies $1 \leqslant m_{10} \leqslant 100$. While their bolometric intensity is low, these particles are capable of significant EBL contributions in narrow portions of the gamma-ray background. To keep the diagram from becoming too cluttered, we have assumed values of $f_\tau$ such that the highest predicted intensity in each case stays just below the EGRET limits. Numerically, this corresponds to lower bounds on the decay lifetime $\tau_{\tilde\chi}$ of between 100 Gyr (for $m_{\tilde\chi}c^2 = 10$ GeV) and $10^5$ Gyr (for $m_{\tilde\chi}c^2 = 300$ GeV). For rest energies at the upper end of this range, these limits are probably optimistic because the decay photons are energetic enough to undergo pair production on CMB photons. Some would be re-processed into lower energies before reaching us. As we show in the next section, however, stronger limits arise from a different process in any case. We defer further discussion of Fig. 8.4 to Sec. 8.6.

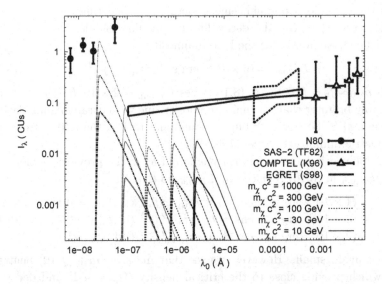

Fig. 8.4    The spectral intensity of the diffuse gamma-ray background due to neutralino one-loop decays (lower left), compared with observational upper limits from high-altitude balloon experiments (filled dots), SAS-2, EGRET and COMPTEL. The three plotted curves for each value of $m_{\tilde{\chi}}c^2$ correspond to models with $\Omega_{m,0} = 0.1h_0$ (heavy lines), $\Omega_{m,0} = 0.3$ (medium lines) and $\Omega_{m,0} = 1$ (light lines). For clarity we have assumed decay lifetimes in each case such that highest theoretical intensities lie just under the observational constraints.

## 8.4    Tree-level decay

The dominant decay processes for the LSP neutralino in non-minimal SUSY (assuming spontaneously broken R-parity) are *tree-level* decays to leptons and neutrinos, $\tilde{\chi} \rightarrow \ell^+ + \ell^- + \nu_\ell$. Of particular interest is the case $\ell = e$, depicted in Fig. 8.5. Although these processes do not contribute directly to the EBL, they do so indirectly, because the high-energy electrons undergo inverse Compton scattering (ICS) off the CMB photons via $e + \gamma_{cmb} \rightarrow e + \gamma$. This gives rise to a flux of high-energy photons which can be at least as important as that from the direct (one-loop) neutralino decays considered in the previous section [393].

The spectrum of photons produced in this way depends on the rest energy of the original neutralino. We consider first the case $m_{10} \lesssim 10$, which is more or less pure ICS. The input ("zero-generation") electrons are monoenergetic, but after multiple scatterings they are distributed like $E^{-2}$

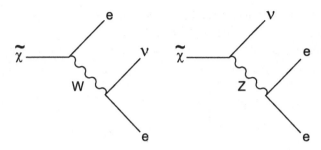

Fig. 8.5   Feynman diagrams corresponding to tree-level decays of the neutralino ($\tilde{\chi}$) into a neutrino ($\nu$) and a lepton-antilepton pair (here, the electron and positron). The process can be mediated by the W or Z-boson.

[394]. From this, the spectrum of outgoing photons can be calculated [395]:

$$N_{\rm ics}(E) \propto \begin{cases} E^{-3/2} & (E \leqslant E_{\rm max}) \\ 0 & (E > E_{\rm max}) \end{cases} , \qquad (8.19)$$

where

$$E_{\rm max} = \frac{4}{3} \left( \frac{E_e}{m_e c^2} \right)^2 E_{\rm cmb} .$$

Here $E_e = \frac{1}{3} m_{\tilde{\chi}} c^2 = (3.3 \text{ GeV}) \, m_{10}$ is the energy of the input electrons, $m_e$ is their rest mass, and $E_{\rm cmb} = 2.7 k T_{\rm cmb}$ is the mean energy of the CMB photons. Using $m_e c^2 = 0.51$ MeV and $T_{\rm cmb} = 2.7$ K, and allowing for decays at arbitrary redshift $z$ [395], we obtain the expression $E_{\rm max}(z) = (36 \text{ keV}) \, m_{10}^2 (1 + z)^{-1}$.

The halo SED may be determined as a function of wavelength by setting $F(\lambda) d\lambda = E N(E) dE$ where $E = hc/\lambda$. Normalizing the spectrum so that $\int_0^\infty F(\lambda) d\lambda = L_{h,\rm tree}$, we find:

$$F_{\rm ics}(\lambda) = \frac{L_{h,\rm tree}}{2} \times \begin{cases} \sqrt{\lambda_\gamma} \, \lambda^{-3/2} & (\lambda \geqslant \lambda_\gamma) \\ 0 & (\lambda < \lambda_\gamma) \end{cases} , \qquad (8.20)$$

where $\lambda_\gamma = hc/E_{\rm max} = (0.34 \text{ Å}) \, m_{10}^{-2} (1 + z)$ and $L_{h,\rm tree}$ is the halo luminosity due to tree-level decays.

In the case of more massive neutralinos with $m_{10} \gtrsim 10$, the situation is complicated by the fact that outgoing photons become energetic enough to initiate pair production via $\gamma + \gamma_{\rm cmb} \rightarrow e^+ + e^-$. This injects new electrons into the ICS process, resulting in electromagnetic cascades. For particles which decay at high redshifts ($z \gtrsim 100$), other processes such as photon-photon scattering must also be taken into account [396]. Cascades on *non-*CMB background photons may also be important [397]. A full treatment of these effects requires detailed numerical analysis [398]. Here we simplify the problem by assuming that the LSP is stable enough to survive into the late matter-dominated (or vacuum-dominated) era. The primary effect of cascades is to steepen the decay spectrum at high energies, so that [395]:

$$N_{\text{casc}}(E) \propto \begin{cases} E^{-3/2} & (E \leqslant E_x) \\ E^{-2} & (E_x < E \leqslant E_c) \\ 0 & (E > E_c) \end{cases} , \qquad (8.21)$$

where

$$E_x = \frac{1}{3}\left(\frac{E_0}{m_e c^2}\right)^2 E_{\text{cmb}}(1+z)^{-1} \qquad E_c = E_0(1+z)^{-1} .$$

Here $E_0$ is a minimum absorption energy. Numerical work has shown that $E_x = (1.8 \times 10^3 \text{ GeV})(1+z)^{-1}$ and $E_c = (4.5 \times 10^4 \text{ GeV})(1+z)^{-1}$ [399]. Employing the relation $F(\lambda)d\lambda = EN(E)dE$ and normalizing as before, we find:

$$F_{\text{casc}}(\lambda) = \frac{L_{h,\text{tree}}}{[2 + \ln(\lambda_x/\lambda_c)]} \times \begin{cases} \sqrt{\lambda_x}\,\lambda^{-3/2} & (\lambda \geqslant \lambda_x) \\ \lambda^{-1} & (\lambda_x > \lambda \geqslant \lambda_c) \\ 0 & (\lambda < \lambda_c) \end{cases} , \quad (8.22)$$

where the new parameters are $\lambda_x = hc/E_x = (7 \times 10^{-9} \text{ Å})(1+z)$ and $\lambda_c = hc/E_c = (3 \times 10^{-10} \text{ Å})(1+z)$.

The luminosity $L_{h,\text{tree}}$ is given by

$$L_{h,\text{tree}} = \frac{N_{\tilde{\chi}} b_e E_e}{\tau_{\tilde{\chi}}} = \frac{2\,b_e M_h c^2}{3\tau_{\tilde{\chi}}} , \qquad (8.23)$$

where $b_e$ is now the branching ratio for all processes of the form $\tilde{\chi} \to e+$ all, and $E_e = \frac{2}{3}m_{\tilde{\chi}}c^2$ is the total energy lost to the electrons. We assume that all of this eventually finds its way into the EBL. Berezinsky *et al.* [392] estimate this branching ratio as

$$b_e \approx 10^{-6} f_\chi f_{\text{R}}^2 m_{10}^2 . \qquad (8.24)$$

Here $f_\chi$ parametrizes the composition of the neutralino, taking the value 0.4 for the pure higgsino case. With the halo mass specified by (6.13) and $f_\tau \equiv \tau_{\tilde{\chi}}/(1 \text{ Gyr})$ as usual, we obtain:

$$L_{h,\text{tree}} = (8 \times 10^{43} \text{ erg s}^{-1})\, m_{10}^2 f_\chi f_{\text{R}}^2 f_\tau^{-1} . \qquad (8.25)$$

This is approximately four orders of magnitude higher than the halo luminosity due to one-loop decays, and provides for the first time the possibility of significant EBL contributions. With all adjustable parameters taking values of order unity, we find that $L_{h,\text{tree}} \sim 2 \times 10^{10} L_\odot$, which is comparable to the bolometric luminosity of the Milky Way.

The combined bolometric intensity of all neutralino halos is computed just as in the previous two sections. Replacing the quantity $L_{h,\text{loop}}$ in Eq. (8.10) with $L_{h,\text{tree}}$ gives

$$Q = \begin{cases} (2 \times 10^{-4} \text{ erg s}^{-1} \text{ cm}^{-2}) \, h_0^2 \, m_{10}^2 f_\chi f_R^2 f_\tau^{-1} \text{ (if } \Omega_{m,0} = 0.1 h_0) \\ (5 \times 10^{-4} \text{ erg s}^{-1} \text{ cm}^{-2}) \, h_0 \, m_{10}^2 f_\chi f_R^2 f_\tau^{-1} \text{ (if } \Omega_{m,0} = 0.3) \\ (2 \times 10^{-3} \text{ erg s}^{-1} \text{ cm}^{-2}) \, h_0 \, m_{10}^2 f_\chi f_R^2 f_\tau^{-1} \text{ (if } \Omega_{m,0} = 1) \end{cases}$$

(8.26)

These results are of the same order as (or higher than) the bolometric intensity of the EBL from ordinary galaxies, Eq. (2.19).

To obtain the spectral intensity, we substitute the SEDs $F_{\text{ics}}(\lambda)$ and $F_{\text{casc}}(\lambda)$ into Eq. (3.6). The results can be written

$$I_\lambda(\lambda_0) = I_{\tilde{\chi},\text{tree}} \int_0^{z_f} \frac{\mathcal{F}(z) \, dz}{(1+z) \, [\Omega_{m,0}(1+z)^3 + (1-\Omega_{m,0})]^{1/2}} \,, \qquad (8.27)$$

where the quantities $I_{\tilde{\chi},\text{tree}}$ and $\mathcal{F}(z)$ are defined as follows. For neutralino rest energies $m_{10} \lesssim 10$ (ICS):

$$I_{\tilde{\chi},\text{tree}} = \frac{n_0 \, L_{h,\text{tree}} \, f_c}{8\pi h \, H_0 \, m_{10}} \left( \frac{0.34 \text{ Å}}{\lambda_0} \right)^{1/2}$$

$$= (300 \text{ CUs}) \, h_0^2 \, m_{10} \, f_\chi \, f_R^2 \, f_\tau^{-1} f_c \left( \frac{\lambda_0}{\text{Å}} \right)^{-\frac{1}{2}} \qquad (8.28)$$

$$\mathcal{F}(z) = \begin{cases} 1 & [\lambda_0 \geqslant \lambda_\gamma(1+z)] \\ 0 & [\lambda_0 < \lambda_\gamma(1+z)] \end{cases} .$$

Conversely, for $m_{10} \gtrsim 10$ (cascades):

$$I_{\tilde{\chi},\text{tree}} = \frac{n_0 \, L_{h,\text{tree}} \, f_c}{4\pi h \, H_0 \, [2 + \ln(\lambda_x/\lambda_c)]} \left( \frac{7 \times 10^{-9} \text{ Å}}{\lambda_0} \right)^{1/2}$$

$$= (0.02 \text{ CUs}) \, h_0^2 \, m_{10}^2 \, f_\chi \, f_R^2 \, f_\tau^{-1} f_c \left( \frac{\lambda_0}{\text{Å}} \right)^{-\frac{1}{2}} \qquad (8.29)$$

$$\mathcal{F}(z) = \begin{cases} 1 & [\lambda_0 \geqslant \lambda_x(1+z)] \\ \dfrac{\lambda_0}{\lambda_x(1+z)} & [\lambda_x(1+z) > \lambda_0 \geqslant \lambda_c(1+z)] \\ 0 & [\lambda_0 < \lambda_c(1+z)] \end{cases} .$$

Numerical integration of Eq. (8.27) leads to the plots in Fig. 8.6. Cascades (like the pair annihilations we have considered already) dominate the gamma-ray part of the spectrum. The ICS process, however, is most important at lower energies, in the x-ray region. We discuss the observational limits and the constraints that can be drawn from them in more detail in Sec. 8.6.

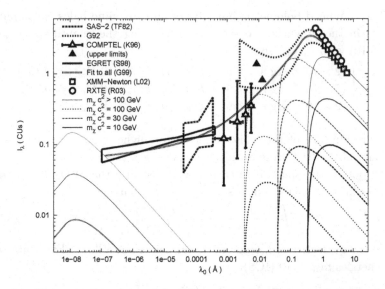

Fig. 8.6   The spectral intensity of the diffuse gamma-ray and x-ray backgrounds due
to neutralino tree-level decays, compared with observational upper limits from SAS-2,
EGRET and COMPTEL in the gamma-ray region, from XMM-Newton and RXTE in the
x-ray region, and from Gruber's fits to various experimental data (G92,G99). The three
plotted curves for each value of $m_{\tilde\chi}c^2$ correspond to models with $\Omega_{m,0} = 0.1h_0$ (heavy
lines), $\Omega_{m,0} = 0.3$ (medium lines) and $\Omega_{m,0} = 1$ (light lines). For clarity we have
assumed decay lifetimes in each case such that the highest theoretical intensities lie just
under the observational constraints.

## 8.5   Gravitinos

Gravitinos ($\tilde{g}$) are the SUSY spin-$\frac{3}{2}$ counterparts of gravitons. Although
often mentioned as dark-matter candidates along with neutralinos, they are
not favored in the simplest SUSY theories. The reason for this, known as
the *gravitino problem* [337], boils down to the fact that they interact *too*
weakly, not only with other particles but with themselves as well. Hence
they annihilate slowly and survive long enough to "overclose" the Universe
unless some other way is found to reduce their numbers. Decays are one
possibility, but not if the gravitino is a stable LSP. Gravitino decay products
must also not be allowed to interfere with processes such as big-bang nucleosynthesis [336]. Inflation, followed by a judicious period of reheating, can
thin out their numbers to almost any desired level. But the reheat temperature $T_R$ must satisfy $kT_R \lesssim 10^{12}$ GeV or gravitinos will once again become
too numerous [341]. Related arguments based on entropy production, nu-

cleosynthesis and the power spectrum of the CMB force this number down to $kT_R \lesssim (10^9 - 10^{10})$ GeV [400] or even $kT_R \lesssim (10^6 - 10^9)$ GeV [401]. These temperatures are incompatible with the generation of baryon asymmetry in the Universe, a process which is usually taken to require $kT_R \sim 10^{14}$ GeV or higher [129].

Recent developments are however beginning to loosen the baryogenesis requirement [402], and there are alternative models in which baryon asymmetry is generated at energies as low as $\sim 10$ TeV [403] or even 10 MeV – 1 GeV [404]. With this in mind we include a brief look at gravitinos here. There are two possibilities: (1) If the gravitino is *not* the LSP, then it decays early in the history of the Universe, well before the onset of the matter-dominated era. In models where the gravitino decays both radiatively and hadronically, for example, it can be "long-lived for its mass" with a lifetime of $\tau_{\tilde{g}} \lesssim 10^6$ s [405]. Particles of this kind have important consequences for nucleosynthesis, and might affect the shape of the CMB if $\tau_{\tilde{g}}$ were to exceed $\sim 10^7$ s. However, they are irrelevant as far as the EBL is concerned. We therefore restrict our attention to the case (2), in which the gravitino is the LSP. In light of the results we have already obtained for the neutralino, we disregard annihilations and consider only models in which the LSP can decay.

The decay mode depends on the specific mechanism of R-parity violation. We follow Berezinsky [406] and concentrate on dominant *tree-level* processes. In particular we consider the decay $\tilde{g} \to e^+ +$ all , followed by ICS off the CMB, as in Sec.8.4. The spectrum of photons produced by this process is identical to that in the neutralino case, except that the mono-energetic electrons have energy $E_e = \frac{1}{2}m_{\tilde{g}}c^2 = (5 \text{ GeV}) m_{10}$ [406], where $m_{\tilde{g}}c^2$ is the rest energy of the gravitino and $m_{10} \equiv m_{\tilde{g}}c^2/(10 \text{ GeV})$ as before. This in turn implies that $E_{\max} = (81 \text{ keV}) m_{10}^2 (1 + z)^{-1}$ and $\lambda_\gamma = hc/E_{max} = (0.15 \text{ Å}) m_{10}^{-2}(1 + z)$. The values of $\lambda_x$ and $\lambda_c$ are unchanged.

The SED comprises Eqs. (8.20) for ICS and (8.22) for cascades, as before. Only the halo luminosity needs to be recalculated. This is similar to Eq. (8.23) for neutralinos, except that the factor of $\frac{2}{3}$ becomes $\frac{1}{2}$, and the branching ratio can be estimated at [406]

$$b_e \sim (\alpha/\pi)^2 = 5 \times 10^{-6} \ . \tag{8.30}$$

Using our standard value for the halo mass $M_h$, and parametrizing the gravitino lifetime by $f_\tau \equiv \tau_{\tilde{g}}/(1 \text{ Gyr})$ as before, we obtain the following halo luminosity due to gravitino decays:

$$L_{h,\text{grav}} = (3 \times 10^{44} \text{ erg s}^{-1}) f_\tau^{-1} \ . \tag{8.31}$$

This is higher than the luminosity due to neutralino decays, and exceeds the luminosity of the Milky Way by several times if $f_\tau \sim 1$.

The bolometric intensity of all gravitino halos is computed exactly as before. Replacing $L_{h,\text{tree}}$ in (8.10) with $L_{h,\text{grav}}$, we find:

$$Q = \begin{cases} (7 \times 10^{-4} \text{ erg s}^{-1} \text{ cm}^{-2})\, h_0^2\, f_\tau^{-1} & (\text{if } \Omega_{m,0} = 0.1h_0) \\ (2 \times 10^{-3} \text{ erg s}^{-1} \text{ cm}^{-2})\, h_0\, f_\tau^{-1} & (\text{if } \Omega_{m,0} = 0.3) \\ (8 \times 10^{-3} \text{ erg s}^{-1} \text{ cm}^{-2})\, h_0\, f_\tau^{-1} & (\text{if } \Omega_{m,0} = 1) \end{cases} \qquad (8.32)$$

It is clear that gravitinos must decay on timescales longer than the lifetime of the Universe ($f_\tau \gtrsim 16$), or they would produce a background brighter than that of the galaxies.

The spectral intensity is the same as before, Eq. (8.27), but with the new numbers for $\lambda_\gamma$ and $L_h$. This results in

$$I_\lambda(\lambda_0) = I_{\tilde{g}} \int_0^{z_f} \frac{\mathcal{F}(z)\, dz}{(1+z)\, [\Omega_{m,0}(1+z)^3 + (1-\Omega_{m,0})]^{1/2}}, \qquad (8.33)$$

where the prefactor $I_{\tilde{g}}$ is defined as follows. For $m_{10} \lesssim 10$ (ICS):

$$I_{\tilde{g}} = \frac{n_0\, L_{h,\text{grav}}\, f_c}{8\pi h\, H_0\, m_{10}} \left( \frac{0.15\text{ Å}}{\lambda_0} \right)^{1/2}$$

$$= (800 \text{ CUs})\, h_0^2\, m_{10}^{-1}\, f_\tau^{-1} f_c \left( \frac{\lambda_0}{\text{Å}} \right)^{-\frac{1}{2}}. \qquad (8.34)$$

Conversely, for $m_{10} \gtrsim 10$ (cascades):

$$I_{\tilde{g}} = \frac{n_0\, L_{h,\text{grav}}\, f_c}{4\pi h\, H_0\, [2 + \ln(\lambda_x/\lambda_c)]} \left( \frac{7 \times 10^{-9}\text{ Å}}{\lambda_0} \right)^{1/2}$$

$$= (0.06 \text{ CUs})\, h_0^2\, f_\tau^{-1} f_c \left( \frac{\lambda_0}{\text{Å}} \right)^{-\frac{1}{2}}. \qquad (8.35)$$

The function $\mathcal{F}(z)$ has the same form as in Eqs. (8.28) and (8.29) and does not need to be redefined (requiring only the new value for the cutoff wavelength $\lambda_\gamma$). Because the branching ratio $b_e$ in (8.30) is independent of the gravitino rest mass, $m_{10}$ appears in these results only through $\lambda_\gamma$. Thus the ICS part of the spectrum goes as $m_{10}^{-1}$ while the cascade part does not depend on $m_{10}$ at all. As with neutralinos, cascades dominate the gamma-ray part of the spectrum, and the ICS process is most important in the x-ray region. Numerical integration of Eq. (8.33) leads to the results plotted in Fig. 8.7. We discuss these in the next section, beginning with an overview of the observational constraints.

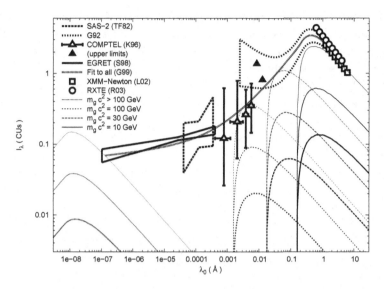

Fig. 8.7 The spectral intensity of the diffuse gamma-ray and x-ray backgrounds due to gravitino tree-level decays, compared with experimental data from SAS-2, EGRET, COMPTEL, XMM-Newton and RXTE, as well as compilations by Gruber. The three plotted curves for each value of $m_{\tilde{\chi}}c^2$ correspond to models with $\Omega_{m,0} = 0.1h_0$ (heavy lines), $\Omega_{m,0} = 0.3$ (medium lines) and $\Omega_{m,0} = 1$ (light lines). For clarity we have assumed decay lifetimes in each case such that the highest theoretical intensities lie just under the observational constraints.

## 8.6 WIMPs and the background light

The experimental situation as regards EBL intensity in the x-ray and gamma-ray regions is more settled than that in the optical and ultraviolet. Detections (as opposed to upper limits) have been made in both bands, and are consistent with expectations based on known astrophysical sources. New measurements have recently been reported from Chandra at the lowest or soft x-ray energies, which lie roughly between 0.1–3 keV (4–100 Å) [407]. (Conventions have not been established regarding the boundaries between different wavebands; we follow most authors and define these according to the different detection techniques that must be used in each region.) These data appear as a small bowtie-shaped box near $\lambda_0 \sim 10$ Å in Fig. 1.15, where it can be seen that they interpolate smoothly between previous detections at shorter and longer wavelengths. The small rectangle immediately to the right of the Chandra bowtie (near $\lambda_0 \sim 100$ Å) in Fig. 1.15 comes from measurements by the EUVE satellite in 1993 [408].

The hard x-ray background (3–800 keV, or 0.02–4 Å) band is crucial in constraining the decays of low-mass neutralinos and gravitinos via the ICS process, as can be seen in Figs. 8.6 and 8.7. We have plotted two compilations of observational data in the hard x-ray band, both by D.E. Gruber [409, 410]. The first (labelled "G92" in Figs. 8.6 and 8.7) is an empirical fit to various pre-1992 measurements, including those from the Kosmos and Apollo spacecraft, HEAO-1 and balloon experiments. The range of uncertainty in these data increases logarithmically from 2% at 3 keV to 60% at 3 MeV [409]. The second compilation (labelled "G99") is a revision of this fit in light of new data at higher energies, and has been extended deep into the gamma-ray region. New results from XMM-Newton [411] and RXTE [412] confirm the accuracy of this revised fit at low energies ("L02" and "R03" respectively in Figs. 8.6 and 8.7). The prominent peak in the range 3–300 keV (0.04–4 Å) is widely attributed to integrated light from active galactic nuclei (AGN) [413].

In the low-energy gamma-ray region (0.8–30 MeV, or 0.0004–0.02 Å) we have used results from the COMPTEL instrument on the Compton Gamma-Ray Observatory (CGRO), which was operational from 1990-2000 [414]. Four data points are plotted in Figs. 8.2, 8.4, 8.6 and 8.7, and two more (upper limits only) appear at low energies in Figs. 8.6 and 8.7. These experimental results, which interpolate smoothly between other data at both lower and higher energies, played a key role in the demise of the "MeV bump" (visible in Figs. 8.6 and 8.7 as a significant upturn in Gruber's fit to the pre-1992 data from about 0.002–0.02 Å). This apparent feature in the background had attracted a great deal of attention from theoretical cosmologists as a possible signature of new physics. Figs. 8.6 and 8.7 suggest that it could also have been interpreted as the signature of a long-lived non-minimal SUSY WIMP with a rest energy near 100 GeV. The MeV bump is, however, no longer believed to be real, as the new fit ("G99") makes clear. Most of the background in this region is now suspected to be due to Type Ia supernovae (SNIa) [415].

We have included two measurements in the high-energy gamma-ray band (30 MeV–30 GeV, or $4 \times 10^{-7}$–$4 \times 10^{-4}$ Å): one from the SAS-2 satellite which flew in 1972-3 [416] and one from the EGRET instrument which was part of the CGRO mission along with COMPTEL [417]. As may be seen in Figs. 8.2, 8.4, 8.6 and 8.7, the new results essentially extend the old ones to 120 GeV ($\lambda_0 = 10^{-7}$ Å), with error bars which have been reduced by a factor of about ten. Most of this extragalactic background is thought to arise from unresolved blazars, which are probably highly-variable AGN whose relativistic jets point in our direction [418]. Some authors have recently argued that Galactic contributions to the background were underestimated in the original EGRET analysis [419, 420]. If so, the true extragalactic background intensity would be lower than that plotted here, strengthening the constraints we derive below.

Because the extragalactic component of the gamma-ray background has not been reliably detected beyond 120 GeV, we have fallen back on measurements of *total* flux in the very high-energy (VHE) region (30 GeV–30 TeV, or $4 \times 10^{-10}$–$4 \times 10^{-7}$ Å). These were obtained from a series of balloon experiments by J. Nishimura *et al.* in 1980 [421], and appear in Figs. 8.2 and 8.4 as filled dots (labelled "N80"). They constitute a very robust upper limit on EBL flux, since much of this signal must have originated in the upper atmosphere. At the very highest energies, in the ultra high-energy (UHE) region ($> 30$ TeV), these data join smoothly to upper limits on the diffuse gamma-ray flux from extensive air-shower arrays such as HEGRA (20-100 TeV [422]) and CASA-MIA (330 TeV-33 PeV [423]). Here we reach the edge of the EBL for practical purposes, since gamma-rays with energies of $\sim 10 - 100$ PeV (1 PeV $\equiv 10^{15}$ eV) are attenuated by pair production on CMB photons over scales $\sim 30$ kpc [424].

Some comments are in order here about units. For experimental reasons, measurements of x-ray and gamma-ray backgrounds are often expressed in terms of integral flux $EI_E(> E_0)$, or number of photons with energies above $E_0$. This presents no difficulties since the differential spectrum in this region is well approximated with a single power-law component, $I_E(E_0) = I_*(E_0/E_*)^{-\alpha}$. The conversion to integral form is then given by

$$EI_E(> E_0) = \int_{E_0}^{\infty} I_E(E) \, dE = \frac{E_* I_*}{\alpha - 1} \left( \frac{E_0}{E_*} \right)^{1-\alpha} . \qquad (8.36)$$

The spectrum is specified in either case by its index $\alpha$ together with the values of $E_*$ and $I_*$ (or $E_0$ and $EI_E$ in the integral case). Thus SAS-2 results were reported as $\alpha = 2.35^{+0.4}_{-0.3}$ with $EI_E = (5.5 \pm 1.3) \times 10^{-5}$ s$^{-1}$ cm$^{-2}$ ster$^{-1}$ for $E_0 = 100$ MeV [416]. The EGRET spectrum is instead fit by $\alpha = 2.10 \pm 0.03$ with $I_* = (7.32 \pm 0.34) \times 10^{-9}$ s$^{-1}$ cm$^{-2}$ ster$^{-1}$ MeV$^{-1}$ for $E_* = 451$ MeV [417]. To convert a differential flux in these units to $I_\lambda$ in CUs, one multiplies by $E_0/\lambda_0 = E_0^2/hc = 80.66 E_0^2$ where $E_0$ is photon energy in MeV.

We now discuss our results, beginning with the neutralino annihilation fluxes plotted in Fig. 8.2. These are at least four orders of magnitude fainter than the background detected by EGRET [417] (and five orders of magnitude below the upper limit set by the data of Nishimura *et al.* [421] at shorter wavelengths). This agrees with previous studies assuming a critical density of neutralinos [354, 383]. Fig. 8.2 shows that EBL contributions would drop by another order of magnitude in the favored scenario with $\Omega_{m,0} \approx 0.3$, and by another if neutralinos are confined to galaxy halos ($\Omega_{m,0} \approx 0.1h_0$). Because the annihilation rate goes as the square of the WIMP density, it has been argued that modelling WIMP halos with sharp density cusps might raise their luminosity, possibly enhancing their EBL contributions by a factor of as much as $10^4 - 10^5$ [425]. While such a scenario might in principle bring WIMP annihilations up to the brink of observability in the diffuse background, density profiles with the required steepness are not seen in either our own Galaxy or those nearby. More recent assessments have reconfirmed the general outlook discussed above in Sec. 8.2; namely, that the best place to look for WIMP annihilations is in the direction of nearby concentrations of dark matter, such as the Galactic center and dwarf spheroidal galaxies in the Local Group [372, 373]. The same stability that makes minimal-SUSY WIMPs so compelling as dark-matter candidates also makes them hard to detect.

Fig. 8.4 shows the EBL contributions from one-loop neutralino decays in *non*-minimal SUSY. We have put $h_0 = 0.75$, $z_f = 30$ and $f_R = 1$. Depending on their decay lifetime (here parametrized by $f_\tau$), these particles are capable in principle of producing a backgound comparable to (or even in excess of) the EGRET limits. The plots in Fig. 8.4 correspond to the smallest values of $f_\tau$ that are consistent with the data for $m_{10} = 1, 3, 10, 30$ and 100. Following the same procedure here as we did for axions in Chap. 6, we can repeat this calculation over more finely-spaced intervals in neutralino rest mass, obtaining a lower limit on decay lifetime $\tau_{\tilde{\chi}}$ as a function of $m_{\tilde{\chi}}$. Results are shown in Fig. 8.8 (dotted lines). The lower limit obtained in this way varies from 4 Gyr for the lightest neutralinos (assumed to be confined to galaxy halos with a total matter density of $\Omega_{m,0} = 0.1h_0$) to 70,000 Gyr for the heaviest (which provide enough CDM to put $\Omega_{m,0} = 1$).

Fig. 8.6 is a plot of EBL flux from *indirect* neutralino decays via the tree-level, ICS and cascade processes described in Sec. 8.4. These provide us with our strongest constraints on non-minimal SUSY WIMPs. We have set $h_0 = 0.75$, $z_f = 30$ and $f_\chi = f_R = 1$, and assumed values of $f_\tau$ such that the highest predicted intensities lie just under observational limits, as before. Neutralinos at the light end of the mass range are constrained by x-ray data, while those at the heavy end run up against the EGRET

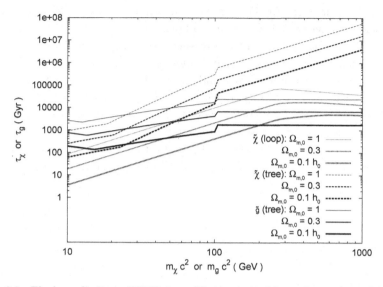

Fig. 8.8    The lower limits on WIMP decay lifetime derived from observations of the x-ray and gamma-ray backgrounds. Neutralino bounds are shown for both one-loop decays (dotted lines) and tree-level decays (dashed lines). For gravitinos we show only the tree-level constraints (solid lines). For each process there are three curves corresponding to models with $\Omega_{m,0} = 1$ (light lines), $\Omega_{m,0} = 0.3$ (medium lines) and $\Omega_{m,0} = 0.1 h_0$ (heavy lines).

measurements. Both the shape and absolute intensity of the ICS spectra depend on the neutralino rest mass, but the cascade spectra depend on $m_{10}$ through intensity alone (via the prefactor $I_{\tilde{\chi},\text{tree}}$). When normalized to the observational upper bound, all curves for $m_{10} > 10$ therefore overlap. Normalizing across the full range of neutralino rest masses (as for one-loop decays) gives the lower bound on lifetime $\tau_{\tilde{\chi}}$ plotted in Fig. 8.8 (dashed lines). This varies from 60 Gyr to $6 \times 10^7$ Gyr, depending on the values of $m_{\tilde{\chi}} c^2$ and $\Omega_{m,0}$.

Fig. 8.7, finally, shows the EBL contributions from tree-level *gravitino* decays. These follow the same pattern as the neutralino decays. Requiring that the predicted signal not exceed the x-ray and gamma-ray observations, we obtain the lower limits on decay lifetime plotted in Fig. 8.8 (solid lines). The flatness of these curves (relative to the constraints on neutralinos) is a consequence of the fact that the branching ratio (8.30) is independent of $m_{10}$. Our lower limits on $\tau_{\tilde{g}}$ range from 200 Gyr to 20,000 Gyr.

Let us sum up our findings in this chapter. We have considered neutralinos and gravitinos, either of which could be the LSP and hence make up the dark matter. In the context of non-minimal SUSY theories, these particles can decay and contribute to both the x-ray and gamma-ray backgrounds. We have shown that any such decay must occur on timescales longer than $10^2$–$10^8$ Gyr (for neutralinos) or $10^2$–$10^4$ Gyr (for gravitinos), depending on their rest masses and various theoretical input parameters. These constraints confirm that, whether it is a neutralino or gravitino, *the LSP in non-minimal SUSY theories must be very nearly stable* . Since an "almost-stable" LSP would require that R-party conservation be violated at improbably low levels, our results imply that the SUSY WIMP either exists in the context of *minimal* SUSY theories, or not at all.

Chapter 9

# The High-Energy Gamma-ray Background

## 9.1 Primordial black holes

Black holes might appear to be the unlikeliest sources of background radiation. In fact, however, experimental data on EBL intensity constrain black holes more strongly than any of the other dark-matter candidates we have discussed so far. Before explaining how this comes about, we distinguish between "ordinary" black holes (which form via the gravitational collapse of massive stars at the end of their lives) and *primordial black holes* (PBHs) which may have arisen from the collapse of overdense regions in the early Universe. The existence of the former is very nearly an established fact, while the latter remain hypothetical. However, it is PBHs which are of interest to us as potential dark-matter candidates.

The reason for this is as follows. Ordinary black holes come from baryonic progenitors (i.e. stars) and are hence classified with the baryonic dark matter of the Universe. (They are of course not "baryonic" in other respects, since among other things their baryon number is not defined.) Ordinary black holes are therefore subject to the nucleosynthesis bound (4.3), which limits them to less than 5% of the critical density. PBHs are *not* subject to this bound because they form during the radiation-dominated era, before nucleosynthesis begins. Nothing prevents them from making up most of the density in the Universe. Moreover they constitute *cold* dark matter because their velocities are low. (That is, they collectively obey a dust-like equation of state, even though they might individually be better described as "radiation-like" than baryonic.) PBHs were first proposed as dark-matter candidates by Y.B. Zeldovich and I.D. Novikov in 1966 [426], and by S.W. Hawking in 1971 [427].

Black holes contribute to the EBL via a process discovered by Hawking in 1974 and often called *Hawking evaporation* [428]. Photons cannot escape from inside the black hole, but they are produced at or near the horizon by quantum fluctuations in the surrounding curved spacetime. These give rise to a net flux of particles which propagates outward from a black hole (of mass $M$) at a rate proportional to $M^{-2}$ (with the black-hole mass itself dropping at the same rate). For ordinary, stellar-mass black holes, this process occurs so slowly that contributions to the EBL are insignificant, and the designation "black" remains perfectly appropriate over the lifetime of the Universe. PBHs, however, can in principle have masses far smaller than those of a star, leading to correspondingly higher luminosities. Those with $M \lesssim 10^{15}$ g (the mass of a small asteroid) would in fact evaporate quickly enough to shed all their mass over less than less than the age of the Universe, and would already have expired in a blaze of high-energy photons and other elementary particles.

We will use the PBHs themselves as sources of radiation in what follows, taking them to be distributed homogeneously throughout space. The degree to which they actually cluster in the potential wells of galaxies and galaxy clusters is not important here, since we are concerned with their combined PBH contributions to the diffuse background. Another subtlety must, however, be taken into account. Unlike the dark-matter halos of previous sections, PBHs cover such a wide range of masses (and luminosities) that we can no longer treat all sources the same way. Instead we must define quantities like number density and energy spectrum as functions of *mass* as well as time, and integrate our final results over both parameters.

The first step is to identify the distribution of PBH masses at the time when they formed. There is little prospect of probing the time before nucleosynthesis experimentally, so any theory of PBH formation is necessarily speculative to some degree. However, the scenario requiring the least extrapolation from known physics is one in which PBHs arise via the gravitational collapse of small initial density fluctuations on a standard Robertson-Walker background, with an equation of state with the usual form (2.25), and with initial density fluctuations distributed like

$$\delta = \epsilon (M_i/M_f)^{-n} . \tag{9.1}$$

Here $M_i$ is the initial mass of the PBH, $M_f$ is the mass lying inside the particle horizon (or causally connected Universe) at PBH formation time, and $\epsilon$ is an unknown proportionality constant.

PBH formation under these conditions was originally investigated by B.J. Carr [429], who showed that the process is favored over an extended range of masses only if $n = \frac{2}{3}$. Proceeding on this assumption, he found that the initial mass distribution of PBHs formed with masses between $M_i$ and $M_i + dM_i$ per unit comoving volume is

$$n(M_i)\, dM_i = \rho_f M_f^{-2} \zeta \left(\frac{M_i}{M_f}\right)^{-\beta} dM_i \, , \qquad (9.2)$$

where $\rho_f$ is the mean density at formation time. The parameters $\beta$ and $\zeta$ are formally given by $2(2\gamma - 1)/\gamma$ and $\epsilon \exp[-(\gamma - 1)^2/2\epsilon^2]$ respectively, where $\gamma$ is the equation-of-state parameter as usual. However, in the interests of lifting possible restrictions on conditions prevailing in the early Universe, we follow Carr [430] in treating $\beta$ and $\zeta$ as free parameters, not necessarily related to $\gamma$ and $\epsilon$. Insofar as the early Universe was governed by the equation of state (2.25), $\beta$ takes values between 2 (dust-like or "soft") and 3 (stiff or "hard"), with $\beta = \frac{5}{2}$ corresponding to the most natural situation (i.e. $\gamma = \frac{4}{3}$ as for a non-interacting relativistic gas). We allow $\beta$ to take values as high as 4, corresponding to "superhard" early conditions. The parameter $\zeta$, meanwhile, represents the fraction of the Universe which goes into PBHs of mass $M_f$ at time $t_f$. It is a measure of the initial inhomogeneity of the Universe.

The fact that Eq. (9.2) has no exponential cutoff at high mass is important because it allows us (in principle at least) to obtain a substantial cosmological density of PBHs. Since $2 \leqslant \beta \leqslant 4$, however, the power-law distribution is dominated by PBHs of *low* mass. This is the primary reason why PBHs turn out to be so tightly constrained by data on background radiation. It is the low-mass PBHs whose contributions to the EBL via Hawking evaporation are the strongest.

Much effort has gone into identifying alternative PBH formation mechanisms that could give rise to a mass distribution more narrowly peaked in the right range to provide the requisite CDM density without the unwanted background radiation from the low-mass tail. For example, PBHs might arise from a post-inflationary spectrum of density fluctuations which is not perfectly scale-invariant but has a characteristic length scale of some kind [431]. The parameter $\zeta$ in (9.2) would then depend on the inflationary potential (or analogous quantities). This kind of dependence has been discussed in the context of two-stage inflation [432], extended inflation [433], chaotic inflation [434], "plateau" inflation [435], hybrid inflation [436] and inflation with isocurvature fluctuations [437].

A narrow spectrum of masses might also arise if PBHs formed during a spontaneous phase transition rather than arising from primordial fluctuations. The quark-hadron transition [438], grand unified symmetry-breaking transition [439] and Weinberg-Salam phase transition [440] have all been considered in this regard. The initial mass distribution in each case would be peaked near the horizon mass $M_f$ at transition time. The quark-hadron transition has attracted particular attention because PBH formation would be enhanced by a temporary softening of the equation of state; and because $M_f$ for this case is coincidentally close to $M_\odot$, so that PBHs formed at this time might be responsible for MACHO observations of microlensing in the halo [441]. Cosmic string loops have also been explored as possible seeds for PBHs with a peaked mass spectrum [442, 443]. Considerable interest has recently been generated by the discovery that PBHs could provide a physical realization of the theoretical phenomenon known as critical collapse [444]. If this is so, then initial PBH masses would no longer necessarily be clustered near $M_f$.

While any of the above proposals can in principle concentrate the PBH population within a narrow mass range, all of them face the same problem of *fine-tuning* if they are to produce the desired present-day density of PBHs. In the case of inflationary mechanisms, it is the form of the potential which must be adjusted. In others it is the bubble nucleation rate, the string mass per unit length, or the fraction of the Universe going into PBHs at formation time. Thus, while modifications of the initial mass distribution may weaken the "standard" constraints on PBH properties (which we derive below), they do not as yet have a compelling physical basis. Similar comments apply to attempts to link PBHs with specific observational phenomena. It has been suggested, for instance, that PBHs with the right mass could be responsible for certain classes of gamma-ray bursts [445–447], or for long-term quasar variability via microlensing [448, 449]. Other possible connections have been made to diffuse gamma-ray emission from the Galactic halo [450, 451] and the MACHO microlensing events [452, 453]. All of these suggestions, while intriguing, beg the question: "Why this particular mass scale?"

## 9.2   Evolution and density

In order to obtain the comoving number density of PBHs from their initial mass distribution, we use the fact that PBHs evaporate at a rate which is inversely proportional to the square of their masses:

$$\frac{dM}{dt} = -\frac{\alpha}{M^2} \, . \tag{9.3}$$

This applies to uncharged, non-rotating black holes, which is a reasonable approximation in the case of PBHs since these objects radiate away electric charge on timescales shorter than their lifetimes [454], and also give up angular momentum by preferentially emitting particles with spin [455]. The parameter $\alpha$ depends in general on PBH mass $M$, and its behavior was worked out in detail by D.N. Page in 1976 [456]. The most important PBHs are those with $4.5 \times 10^{14}$ g $\leqslant M \leqslant 9.4 \times 10^{16}$ g. Black holes in this range are light (and therefore "hot") enough to emit massless particles (including photons), as well as ultra-relativistic electrons and positrons. The corresponding value of $\alpha$ is

$$\alpha = 6.9 \times 10^{25} \text{ g}^3 \text{ s}^{-1} \ . \tag{9.4}$$

For $M > 9.4 \times 10^{16}$ g, the value of $\alpha$ drops to $3.8 \times 10^{25}$ g$^3$ s$^{-1}$ because larger black holes are "cooler" and no longer able to emit electrons and positrons. EBL contributions from PBHs of this mass are however of lesser importance because of the shape of the mass distribution.

As the PBH mass drops below $4.5 \times 10^{14}$ g, its energy $kT$ climbs past the rest energies of progressively heavier particles, beginning with muons and pions. As each mass threshold is passed, the PBH is able to emit more particles and the value of $\alpha$ increases further. At temperatures above the quark-hadron transition ($kT \approx 200$ MeV), relativistic quark and gluon jets are likely to be emitted rather than massive elementary particles [457]. These jets subsequently fragment into stable particles, and the photons produced in this way are actually more important (at these energies) than the primary photon flux. The precise behavior of $\alpha$ in this regime depends to some extent on the choice of particle physics. A plot of $\alpha(M)$ for the standard model appears in Ref. [458], with the remark that $\alpha$ climbs to $7.8 \times 10^{26}$ g$^3$ s$^{-1}$ at $kT = 100$ GeV; and that its value would be at least three times higher in supersymmetric extensions of the standard model, where there are many more particle states to be emitted.

As we will shortly see, however, EBL contributions from PBHs at these temperatures are suppressed by the fact that the latter have already evaporated. If PBH evolution is well described by (9.3) with $\alpha = $ constant as given by (9.4), then integration gives

$$M(t) = (M_i^3 - 3\alpha t)^{\frac{1}{3}} \ . \tag{9.5}$$

The lifetime $t_{\text{pbh}}$ of a PBH is found by setting $M(t_{\text{pbh}}) = 0$, giving $t_{\text{pbh}} = M_i^3/3\alpha$. Therefore the initial mass of a PBH which is just disappearing today ($t_{\text{pbh}} = t_0$) is given by

$$M_* = (3\alpha t_0)^{\frac{1}{3}} \ . \tag{9.6}$$

Taking $t_0 = 16$ Gyr and using (9.4) for $\alpha$, we find that $M_* = 4.7 \times 10^{14}$ g. A numerical analysis allowing for changes in the value of $\alpha$ over the full range of PBH masses with $0.06 \leqslant \Omega_{m,0} \leqslant 1$ and $0.4 \leqslant h_0 \leqslant 1$ leads to a slightly larger result [458]:

$$M_* = (5.7 \pm 1.4) \times 10^{14} \text{ g} . \tag{9.7}$$

PBHs with $M \approx M_*$ are evaporating at redshift $z \approx 0$ and consequently dominate the spectrum of EBL contributions. The parameter $M_*$ is therefore of central importance in what follows.

We now obtain the comoving number density of PBHs with masses between $M$ and $M + dM$ at any time $t$. This is the same as the comoving number density of PBHs with *initial* masses between $M_i$ and $M_i + dM_i$ at formation time, so $n(M, t) \, dM = n(M_i) \, dM_i$. Inverting Eq. (9.5) to get $M_i = (M^3 + 3\alpha t)^{1/3}$ and differentiating, we find from (9.2) that

$$n(\mathcal{M}, \tau) \, d\mathcal{M} = \mathcal{N} M^2 \left( \mathcal{M}^3 + \tau \right)^{-(\beta+2)/3} d\mathcal{M} . \tag{9.8}$$

Here we have used (9.6) to replace $M_*^3$ with $3\alpha t_0$ and switched to dimensionless parameters $\mathcal{M} \equiv M/M_*$ and $\tau \equiv t/t_0$. The quantity $\mathcal{N}$ is formally given in terms of the parameters at PBH formation time by $\mathcal{N} = (\zeta \, \rho_f/M_f)(M_f/M_*)^{\beta-1}$ and has the dimensions of a number density. As we will see shortly, it corresponds roughly to the comoving number density of PBHs of mass $M_*$. Following Page and Hawking [459], we allow $\mathcal{N}$ to move up or down as required by observational constraints. The theory to this point is thus specified by two free parameters: the PBH normalization $\mathcal{N}$ and the equation-of-state parameter $\beta$.

To convert to the present mass density of PBHs with mass ratios between $\mathcal{M}$ and $\mathcal{M} + d\mathcal{M}$, we multiply Eq. (9.8) by $M = M_* \mathcal{M}$ and put $\tau = 1$:

$$\rho_{\text{pbh}}(\mathcal{M}, 1) \, d\mathcal{M} = \mathcal{N} M_* \mathcal{M}^{1-\beta} \left( 1 + \mathcal{M}^{-3} \right)^{-(\beta+2)/3} d\mathcal{M} . \tag{9.9}$$

The total mass in PBHs is then found by integrating over $\mathcal{M}$ from zero to infinity. Changing the variable to $x \equiv \mathcal{M}^{-3}$, we obtain:

$$\rho_{\text{pbh}} = \frac{1}{3} \mathcal{N} M_* \int_0^\infty x^{a-1} (1 - x)^{-(a+b)} \, dx , \tag{9.10}$$

where $a \equiv \frac{1}{3}(\beta - 2)$ and $b \equiv \frac{4}{3}$. This integral gives

$$\rho_{\text{pbh}} = k_\beta \mathcal{N} M_* \qquad k_\beta \equiv \frac{\Gamma(a)\Gamma(b)}{3\Gamma(a + b)} , \tag{9.11}$$

where $\Gamma(x)$ is the gamma function. Allowing $\beta$ to take values from 2 through $\frac{5}{2}$ (the most natural situation) and up to 4, we find that

$$
k_\beta = \begin{cases}
\infty & \text{(if } \beta = 2) \\
1.87 & \text{(if } \beta = 2.5) \\
0.88 & \text{(if } \beta = 3) \\
0.56 & \text{(if } \beta = 3.5) \\
0.40 & \text{(if } \beta = 4)
\end{cases} \qquad (9.12)
$$

The total mass density of PBHs in the Universe is thus $\rho_{\text{pbh}} \approx \mathcal{N} M_*$. Eq. (9.11) can be recast as a relation between the characteristic number density $\mathcal{N}$ and the PBH density parameter $\Omega_{\text{pbh}} = \rho_{\text{pbh}}/\rho_{\text{crit},0}$:

$$
\Omega_{\text{pbh}} = \frac{k_\beta \mathcal{N} M_*}{\rho_{\text{crit},0}} . \qquad (9.13)
$$

The quantities $\mathcal{N}$ and $\Omega_{\text{pbh}}$ are thus interchangeable as free parameters. If we adopt the most natural value for $\beta$ (=2.5), together with an upper limit on $\mathcal{N}$ due to Page and Hawking of $\mathcal{N} \lesssim 10^4$ pc$^{-3}$ [459], then Eqs. (2.22), (9.7), (9.12) and (9.13) together imply that $\Omega_{\text{pbh}}$ is at most of order $10^{-8} h_0^{-2}$. If this upper limit holds (as we confirm below), then it is unlikely that PBHs make up the dark matter.

Eq. (9.12) shows that one way to boost their importance would be to assume a soft equation of state at formation time. Values of $\beta$ close to two are not ruled out, although Eq. (9.2) becomes invalid in the limit $\beta \to 2$ [429]. Physically, the reason why this produces more PBHs is that low-pressure matter offers little resistance to gravitational collapse. Such a softening has in fact been shown to occur during the quark-hadron transition [441], leading to significant increases in $\Omega_{\text{pbh}}$ for PBHs which form at that time (subject to the fine-tuning problem noted in Sec. 9.1). For PBHs which arise from primordial density fluctuations, however, such conditions are unlikely to hold throughout the formation epoch.

## 9.3   Spectral energy distribution

Hawking [460] proved that an uncharged, non-rotating black hole emits bosons (such as photons) in any given quantum state with energies between $E$ and $E + dE$ at the rate

$$
d\dot{N} = \frac{\Gamma_s \, dE}{2\pi\hbar \left[\exp(E/kT) - 1\right]} . \qquad (9.14)
$$

Here $T$ is the effective black-hole temperature, and $\Gamma_s$ is the absorption coefficient or probability that the same particle would be absorbed by the black hole if incident upon it in this state. The function $d\dot{N}$ is related to the spectral energy distribution (SED) of the black hole by $d\dot{N} = F(\lambda, \mathcal{M}) \, d\lambda / E$, since we have defined $F(\lambda, \mathcal{M}) \, d\lambda$ as the energy emitted between wavelengths $\lambda$ and $\lambda + d\lambda$. We anticipate that $F$ will depend explicitly on the PBH mass $\mathcal{M}$ as well as wavelength. The PBH SED thus satisfies

$$F(\lambda, \mathcal{M}) \, d\lambda = \frac{\Gamma_s E \, dE}{2\pi\hbar \left[\exp(E/kT) - 1\right]} \, . \tag{9.15}$$

The absorption coefficient $\Gamma_s$ is a function of $\mathcal{M}$ and $E$ as well as the quantum numbers $s$ (spin), $\ell$ (total angular momentum) and $m$ (axial angular momentum) of the emitted particles. Its form was first calculated by Page [456]. At high energies, and in the vicinity of the peak of the emitted spectrum, a good approximation is given by [461]:

$$\Gamma_s \propto M^2 E^2 \, . \tag{9.16}$$

This approximation breaks down at low energies, where it gives rise to errors of order 50% for $(GME/\hbar c^3) \sim 0.05$ [462] or (with $E = 2\pi\hbar c/\lambda$ and $M \sim M_*$) for $\lambda \sim 10^{-3}$ Å. This is acceptable for our purposes, as we will find that the strongest constraints on PBHs come from those with masses $M \sim M_*$ at wavelengths $\lambda \sim 10^{-4}$ Å.

Putting (9.16) into (9.15), and making the change of variable to wavelength $\lambda = hc/E$, we obtain the SED

$$F(\lambda, \mathcal{M}) \, d\lambda = \frac{C\mathcal{M}^2 \lambda^{-5} \, d\lambda}{\exp(hc/kT\lambda) - 1} \, , \tag{9.17}$$

where $C$ is a proportionality constant. This has the same form as the Planckian spectrum, Eq. (3.22). We have made three simplifying assumptions in arriving at this result. First, we have neglected the black-hole charge and spin (as justified in Sec. 9.2). Second, we have used an approximation for the absorption coefficient $\Gamma_s$. And third, we have treated all the emitted photons as if they are in the same quantum state, whereas in fact the emission rate (9.14) applies separately to the $\ell = s$ $(= 1)$, $\ell = s + 1$ and $\ell = s + 2$ modes. There are thus actually *three distinct* quasi-blackbody photon spectra with different characteristic temperatures for any single PBH. However, Page [456] has demonstrated that the $\ell = s$ mode is overwhelmingly dominant, with the $\ell = s + 1$ and $\ell = s + 2$ modes contributing less than 1% and 0.01% of the total photon flux respectively. Eq. (9.17) is thus a reasonable approximation for the SED of the PBH as a whole.

To fix the value of $C$ we use the fact that the total flux of photons (in all modes) radiated by a black hole of mass $M$ is given by [456]:

$$\dot{N} = \int d\dot{N} = \int_{\lambda=0}^{\infty} \frac{F(\lambda, \mathcal{M}) \, d\lambda}{hc/\lambda} = 5.97 \times 10^{34} \text{ s}^{-1} \left( \frac{M}{1 \text{ g}} \right)^{-1} . \quad (9.18)$$

Inserting (9.17) and recalling that $M = M_* \mathcal{M}$, we find that

$$C \int_0^{\infty} \frac{\lambda^{-4} \, d\lambda}{\exp(hc/kT\lambda) - 1} = (5.97 \times 10^{34} \text{ g s}^{-1}) \frac{hc}{M_* \mathcal{M}^3} . \quad (9.19)$$

The definite integral on the left-hand side of this equation can be solved by switching variables to $\nu = c/\lambda$:

$$\int_0^{\infty} \frac{\nu^2 \, d\nu / c^3}{\exp(h\nu/kT) - 1} = \left( \frac{hc}{kT} \right)^{-3} \Gamma(3) \, \zeta(3) , \quad (9.20)$$

where $\Gamma(n)$ and $\zeta(n)$ are the gamma function and Riemann zeta function respectively. We then apply the fact that the temperature $T$ of an uncharged, non-rotating black hole is given by

$$T = \frac{\hbar c^3}{8\pi kGM} . \quad (9.21)$$

Putting (9.20) and (9.21) into (9.19) and rearranging terms leads to

$$C = (5.97 \times 10^{34} \text{ g s}^{-1}) \frac{(4\pi)^6 h \, G^3 M_*^2}{c^5 \, \Gamma(3) \, \zeta(3)} . \quad (9.22)$$

Using $\Gamma(3) = 2! = 2$ and $\zeta(3) = 1.202$ along with (9.7) for $M_*$, we find

$$C = (270 \pm 120) \text{ erg s}^{-1} \text{ Å}^4 . \quad (9.23)$$

We can also use the definitions (9.21) to define a useful new quantity:

$$\lambda_{\text{pbh}} \equiv \frac{hc}{kT\mathcal{M}} = \left( \frac{4\pi}{c} \right)^2 GM_* = (6.6 \pm 1.6) \times 10^{-4} \text{ Å} . \quad (9.24)$$

The size of this characteristic wavelength tells us that we will be concerned primarily with the high-energy gamma-ray portion of the spectrum. In terms of $C$ and $\lambda_{\text{pbh}}$ the SED (9.17) now reads

$$F(\lambda, \mathcal{M}) = \frac{C\mathcal{M}^2/\lambda^5}{\exp(\mathcal{M}\lambda_{\text{pbh}}/\lambda) - 1} . \quad (9.25)$$

While this contains no explicit time-dependence, the spectrum does of course depend on time through the PBH mass ratio $\mathcal{M}$. To find the PBH luminosity we employ Eq. (3.1), integrating the SED $F(\lambda, \mathcal{M})$ over all $\lambda$ to obtain:

$$L(\mathcal{M}) = C\mathcal{M}^2 \int_0^{\infty} \frac{\lambda^{-5} \, d\lambda}{\exp(\mathcal{M}\lambda_{\text{pbh}}/\lambda) - 1} . \quad (9.26)$$

This definite integral is also solved by means of a change of variable to frequency $\nu$, with the result that

$$L(\mathcal{M}) = C\mathcal{M}^2 (\mathcal{M}\lambda_{\text{pbh}})^{-4} \Gamma(4) \zeta(4) . \tag{9.27}$$

Using Eqs. (9.7), (9.22) and (9.24) along with the values $\Gamma(4) = 3! = 6$ and $\zeta(4) = \pi^4/96$, we can put this into the form

$$L(\mathcal{M}) = L_{\text{pbh}} \mathcal{M}^{-2} , \tag{9.28}$$

where

$$L_{\text{pbh}} = \frac{(5.97 \times 10^{34} \text{ g s}^{-1}) \pi^2 hc^3}{512\,\zeta(3)\,GM_*^2} = (1.0 \pm 0.4) \times 10^{16} \text{ erg s}^{-1} .$$

Compared to an ordinary star, the typical PBH (of mass ratio $\mathcal{M} \approx 1$) is not very luminous. A PBH of 900 kg or so might theoretically be expected to reach the Sun's luminosity. However, in practice it would already have evaporated, having reached an effective temperature high enough to emit a wide range of massive particles as well as photons. The low luminosity of black holes in general can be emphasized by using the relation $\mathcal{M} \equiv M/M_*$ to recast Eq. (9.28) in the form

$$L/L_\odot = 1.7 \times 10^{-55}(M/M_\odot)^{-2} . \tag{9.29}$$

This expression is not strictly valid for PBHs of masses near $M_\odot$, having been derived for those with $M \sim M_* \sim 10^{15}$ g. (Luminosity is *lower* for larger black holes, and one of solar mass would be so much colder than the CMB that it would absorb radiation faster than it could emit it.) So, Hawking evaporation or not, most black holes are indeed very black.

## 9.4   Bolometric intensity

To obtain the total bolometric intensity of PBHs back to time $t_f$, we substitute the PBH number density (9.8) and luminosity (9.26) into the integral (2.10) as usual. However, now the number density $n(t)$ is to be replaced by $n(\mathcal{M}, \tau)\, d\mathcal{M}$, where $L(\mathcal{M})$ takes the place of $L(t)$, and we integrate over all PBH masses $\mathcal{M}$ as well as times $\tau$. Thus

$$Q = Q_{\text{pbh}}\,\Omega_{\text{pbh}} \int_{\tau_f}^1 \tilde{R}(\tau)\, d\tau \int_0^\infty \frac{d\mathcal{M}}{(\mathcal{M}^3 + \tau)^\varepsilon} , \tag{9.30}$$

where

$$Q_{\text{pbh}} = \frac{ct_0 \rho_{\text{crit},0} L_{\text{pbh}}}{k_\beta M_*} . \tag{9.31}$$

Here $\varepsilon \equiv (\beta + 2)/3$ and we have used (9.13) to replace $\mathcal{N}$ with $\Omega_{\text{pbh}}$. In principle, the integral over $\mathcal{M}$ should be cut off at a finite lower limit $\mathcal{M}_c(\tau)$, equal to the mass of the lightest PBH which has not yet evaporated at time $\tau$. This arises because the initial PBH mass distribution (9.2) requires a finite minimum $M_{\text{min}}$ in order to avoid divergences at low mass. In practice, however, the cutoff rapidly evolves toward zero from its initial value of $\mathcal{M}_c(0) = M_{\text{min}}/M_*$. If $M_{\text{min}}$ is of the order of the Planck mass as usually suggested [463], then $\mathcal{M}_c(\tau)$ drops to zero well before the end of the radiation-dominated era. Since we are concerned with times later than this, we can safely set $\mathcal{M}_c(\tau) = 0$.

Eq. (9.30) can be used to put a rough upper limit on $\Omega_{\text{pbh}}$ from the bolometric intensity of the background light [464]. Let us assume that the Universe is 3D-flat, as suggested by observations (Chap. 4). Then its age $t_0$ can be obtained from Eq. (2.54) as

$$t_0 = 2\,\tilde{t}_0/(3\,H_0) \,. \tag{9.32}$$

Here $\tilde{t}_0 \equiv \tilde{t}_m(0)$ where $\tilde{t}_m(z)$ is the dimensionless function

$$\tilde{t}_m(z) \equiv \frac{1}{\sqrt{1 - \Omega_{m,0}}} \sinh^{-1}\sqrt{\frac{1 - \Omega_{m,0}}{\Omega_{m,0}(1+z)^3}} \,. \tag{9.33}$$

Putting (9.32) into (9.31) and using Eqs. (2.16), (2.22), (9.7) and (9.28), we find:

$$Q_{\text{pbh}} = \frac{(5.97 \times 10^{34} \text{ g s}^{-1})\,\pi^2 h\,c^4 \rho_{\text{crit},0}\,\tilde{t}_0}{768\,\zeta(3)GH_0\,k_\beta M_*^3}$$

$$= (2.3 \pm 1.4)\,h_0\,\tilde{t}_0\,k_\beta^{-1} \text{ erg s}^{-1} \text{ cm}^{-2} \,. \tag{9.34}$$

We are now ready to evaluate Eq. (9.30). To begin with we note that the integral over mass has an analytic solution:

$$\int_0^\infty \frac{d\mathcal{M}}{(\mathcal{M}^3 + \tau)^\varepsilon} = k_\varepsilon \tau^{\frac{1}{3}-\varepsilon} \qquad k_\varepsilon \equiv \frac{\Gamma(\frac{1}{3})\,\Gamma(\varepsilon - \frac{1}{3})}{3\,\Gamma(\varepsilon)} \,. \tag{9.35}$$

For the EdS case ($\Omega_{m,0} = 1$), $\tilde{t}_0 = 1$ and Eq. (2.49) implies:

$$\tilde{R}(\tau) = \tau^{2/3} \,. \tag{9.36}$$

Putting Eqs. (9.35) and (9.36) into (9.30), we find that

$$Q = Q_{\text{pbh}}\,\Omega_{\text{pbh}}\,k_\varepsilon \int_{\tau_f}^1 \tau^{1-\varepsilon}\,d\tau = Q_{\text{pbh}}\,\Omega_{\text{pbh}}\,k_\varepsilon \left(\frac{1 - \tau_f^{2-\varepsilon}}{2 - \varepsilon}\right) \,. \tag{9.37}$$

The parameter $\tau_f$ is obtained for the EdS case by inverting (9.36) to give $\tau_f = (1+z_f)^{-3/2}$. The subscript "$f$" ("formation") is here a misnomer since we do not integrate back to PBH formation time, which occurred in the early stages of the radiation-dominated era. Rather we integrate out to the redshift at which processes like pair production become significant enough to render the Universe approximately opaque to the (primarily gamma-ray) photons from PBH evaporation, $z_f \approx 700$ [462].

Using this value of $z_f$ and substituting Eqs. (9.34) and (9.35) into (9.37), we find that the bolometric intensity of background radiation due to evaporating PBHs in an EdS Universe is

$$
Q = h_0\,\Omega_{\text{pbh}} \times
\begin{cases}
0 & \text{(if } \beta = 2) \\
2.3 \pm 1.4 \text{ erg sec}^{-1}\text{ cm}^{-2} & \text{(if } \beta = 2.5) \\
6.6 \pm 4.2 \text{ erg sec}^{-1}\text{ cm}^{-2} & \text{(if } \beta = 3) \\
17 \pm 10 \text{ erg sec}^{-1}\text{ cm}^{-2} & \text{(if } \beta = 3.5) \\
45 \pm 28 \text{ erg sec}^{-1}\text{ cm}^{-2} & \text{(if } \beta = 4)
\end{cases}
\qquad (9.38)
$$

This vanishes for $\beta = 2$ because $k_\beta \to \infty$ in this limit, as discussed in Sec. 9.2. The case $\beta = 4$ (i.e. $\varepsilon = 2$) is evaluated with the help of L'Hôpital's rule, which gives $\lim_{\varepsilon \to 2}(1 - \tau_f^{2-\varepsilon})/(2 - \varepsilon) = -\ln \tau_f$.

The values of $Q$ in Eq. (9.38) are far higher than the actual bolometric intensity of background radiation in an EdS universe, which from Fig. 2.6 is $1.0 \times 10^{-4}$ erg s$^{-1}$ cm$^{-2}$. Moreover this background is already well accounted for by known astrophysical sources. A firm upper bound on $\Omega_{\text{pbh}}$ (for the most natural situation with $\beta = 2.5$) is therefore

$$
\Omega_{\text{pbh}} < (4.4 \pm 2.8) \times 10^{-5} h_0^{-1} .
\qquad (9.39)
$$

For harder equations of state ($\beta > 2.5$) the PBH density would have to be even lower. PBHs in the simplest formation scenario are thus eliminated as important dark-matter candidates, even without reference to the cosmic gamma-ray background.

For models containing dark energy as well as baryons and black holes, the integrated background intensity goes up because the Universe is older, and down because $Q \propto \Omega_{\text{pbh}}$. The latter effect is stronger, so that the above constraint on $\Omega_{\text{pbh}}$ will be weaker in a model such as $\Lambda$CDM (with $\Omega_{m,0} = 0.3, \Omega_{\Lambda,0} = 0.7$). To determine the importance of this effect, we can re-evaluate the integral (9.30), using the general formula (2.52) for $\tilde{R}(\tau)$ in place of (9.36). We will make the minimal assumption that PBHs constitute the *only* CDM, so that $\Omega_{m,0} = \Omega_{\text{pbh}} + \Omega_{\text{bar}}$ with $\Omega_{\text{bar}}$ given by Eq. (4.3) as usual. Eq. (2.54) shows that the parameter $\tau_f$ is given for arbitrary values of $\Omega_{m,0}$ by $\tau_f = \tilde{t}_m(z_f)/\tilde{t}_0$ where the function $\tilde{t}_m(z)$ is defined as before by Eq. (9.33).

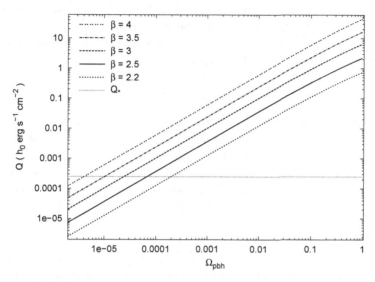

Fig. 9.1   The bolometric intensity due to evaporating primordial black holes as a function of their collective density $\Omega_{\rm pbh}$ and the equation-of-state parameter $\beta$. We have assumed that $\Omega_{m,0} = \Omega_{\rm bar} + \Omega_{\rm pbh}$ with $\Omega_{\rm bar} = 0.016 h_0^{-2}$, $h_0 = 0.75$ and $\Omega_{\Lambda,0} = 1 - \Omega_{m,0}$. The horizontal dotted line indicates the approximate bolometric intensity $(Q_*)$ of the observed EBL.

Evaluation of Eq. (9.30) leads to the plot of bolometric intensity $Q$ versus $\Omega_{\rm pbh}$ in Fig. 9.1. As before, $Q$ is proportional to $h_0$ because it goes as both $\rho_{\rm pbh} = \Omega_{\rm pbh} \rho_{\rm crit,0} \propto h_0^2$ and $t_0 \propto h_0^{-1}$. Since $Q \to 0$ for $\beta \to 2$ we have chosen a minimum value of $\beta = 2.2$ as representative of "soft" conditions. Fig. 9.1 confirms that, regardless of cosmological model, PBH contributions to the background light are too high unless $\Omega_{\rm pbh} \ll 1$. The values in Eq. (9.38) are recovered at the right-hand edge of the figure where $\Omega_{\rm pbh}$ approaches one, as expected. For all other models, we can impose a conservative upper bound $Q < Q_*$ (as indicated in Fig. 9.1 by the horizontal dotted line, corresponding to the observed EBL intensity). Then it follows that $\Omega_{\rm pbh} < (6.9 \pm 4.2) \times 10^{-5} h_0^{-1}$ for $\beta = 2.5$. This is about 60% higher than the limit (9.39) for the EdS case.

## 9.5   Spectral intensity

Stronger limits on PBH density can be obtained from the gamma-ray background, where these objects contribute most strongly to the EBL and where we have good data (as summarized in Sec. 8.6). Spectral intensity is found as usual by substituting the comoving PBH number density (9.8) and SED (9.25) into Eq. (3.5). As in the bolometric case, we now have to integrate over PBH mass $\mathcal{M}$ as well as time $\tau = t/t_0$, so that

$$I_\lambda(\lambda_0) = \frac{ct_0}{4\pi} \int_{\tau_f}^{1} \tilde{R}^2(\tau)\, d\tau \int_{\mathcal{M}_c(\tau)}^{\infty} n(\mathcal{M},\tau)\, F(\tilde{R}\lambda_0, \mathcal{M})\, d\mathcal{M} \, . \qquad (9.40)$$

Following the discussion in Sec. 9.4 we set $\mathcal{M}_c(\tau) = 0$. In light of our bolometric results it is unlikely that PBHs make up a significant part of the dark matter, so we no longer tie the value of $\Omega_{m,0}$ to $\Omega_{\rm pbh}$. Models with $\Omega_{m,0} \gtrsim \Omega_{\rm bar}$ must therefore contain a second species of cold dark matter (other than PBHs) to provide the required matter density. Putting (9.8) and (9.25) into (9.40) and using (9.7), (9.13) and (9.22), we find that

$$I_\lambda(\lambda_0) = I_{\rm pbh}\,\Omega_{\rm pbh} \int_{\tau_f}^{1} \tilde{R}^{-3}(\tau)\, d\tau \int_{0}^{\infty} \frac{\mathcal{M}^4(\mathcal{M}^3 + \tau)^{-\varepsilon}\, d\mathcal{M}}{\exp\left[\lambda_{\rm pbh}\mathcal{M}/\tilde{R}(\tau)\lambda_0\right] - 1} \, . \qquad (9.41)$$

Here the dimensional prefactor is a function of both $\beta$ and $\lambda_0$ and reads

$$I_{\rm pbh} = \frac{(5.97 \times 10^{34}\ {\rm g\ s}^{-1})(4\pi)^5 G^3 M_* \rho_{\rm crit,0}\,\tilde{t}_0}{3\,\zeta(3)\,c^5\,k_\beta\,H_0\,\lambda_0^4}$$

$$= \left[(2.1 \pm 0.5) \times 10^{-7}\ {\rm CUs}\right] h_0\, k_\beta^{-1}\tilde{t}_0 \left(\frac{\lambda_0}{\text{Å}}\right)^{-4} . \qquad (9.42)$$

We have divided through by the photon energy $hc/\lambda_0$ to put the results in units of CUs as usual. The range of uncertainty in $I_\lambda(\lambda_0)$ is smaller than that in $Q$, Eq. (9.34), because $I_\lambda(\lambda_0)$ depends only linearly on $M_*$ whereas $Q$ is proportional to $M_*^{-3}$. (This in turn results from the fact that $I_\lambda \propto C \propto M_*^2$ whereas $Q \propto L_{\rm pbh} \propto M_*^{-2}$. One more factor of $M_*^{-1}$ comes from $\mathcal{N} \propto \rho_{\rm pbh}/M_*$ in both cases.) Like $Q$, $I_\lambda$ depends linearly on $h_0$ since integrated intensity in either case is proportional to both $\rho_{\rm pbh} \propto \rho_{\rm crit,0} \propto h_0^2$ and $t_0 \propto h_0^{-1}$.

Numerical integration of Eq. (9.41) leads to the plots shown in Fig. 9.2, where we have set $\Omega_{\rm pbh} = 10^{-8}$. Following Page and Hawking [459] we have chosen values of $\Omega_{m,0} = 0.06$ in panel (a) and $\Omega_{m,0} = 1$ in panel (b). Our results are in good agreement with the earlier ones except at the longest wavelengths (lowest energies), where PBH evaporation is no longer well described by a simple blackbody SED, and where the spectrum begins to be affected by pair production. As expected, the spectra peak near $10^{-4}$ Å in the gamma-ray region. Also plotted in Fig. 9.2 are the data from SAS-2 ([416]; heavy dashed line), COMPTEL ([414]; triangles) and EGRET ([417]; heavy solid line).

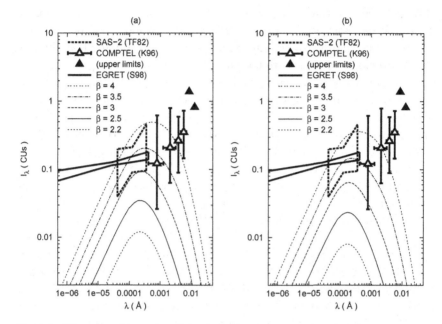

Fig. 9.2 The spectral intensity of the diffuse gamma-ray background from evaporating primordial black holes in flat models, as compared with experimental limits from SAS-2, COMPTEL and EGRET. Panel (a) assumes $\Omega_{m,0} = 0.06$. Panel (b) is plotted for $\Omega_{m,0} = 1$, the EdS case. All curves assume $\Omega_{\rm pbh} = 10^{-8}$ and $h_0 = 0.75$.

By adjusting the value of $\Omega_{\rm pbh}$ up or down from its value of $10^{-8}$ in Fig. 9.2, we can match the theoretical PBH spectra to those measured (e.g., by EGRET), thereby obtaining the maximum value of $\Omega_{\rm pbh}$ consistent with observation. For $\beta = 2.5$ this results in

$$\Omega_{\rm pbh} < \begin{cases} (4.2 \pm 1.1) \times 10^{-8}\, h_0^{-1} & (\text{if } \Omega_{m,0} = 0.06) \\ (6.2 \pm 1.6) \times 10^{-8}\, h_0^{-1} & (\text{if } \Omega_{m,0} = 1) \end{cases} . \tag{9.43}$$

These limits are three orders of magnitude stronger than the one from bolometric intensity, again confirming that PBHs in the simplest formation scenario cannot be significant contributors to the dark matter. Using (9.13) this result can be translated into an upper limit on $\mathcal{N}$:

$$\mathcal{N} < \begin{cases} (2.2 \pm 0.8) \times 10^4\, h_0\ {\rm pc}^{-3} & (\text{if } \Omega_{m,0} = 0.06) \\ (3.2 \pm 1.1) \times 10^4\, h_0\ {\rm pc}^{-3} & (\text{if } \Omega_{m,0} = 1) \end{cases} . \tag{9.44}$$

These numbers are in good agreement with the original Page-Hawking bound of $\mathcal{N} < 1 \times 10^4\ {\rm pc}^{-3}$ [459], which was obtained for $h_0 = 0.6$.

Subsequent workers have refined the gamma-ray background constraints on $\Omega_{\rm pbh}$ and $\mathcal{N}$ in a number of ways. J.H. MacGibbon and B.R. Webber

[457] pointed out that PBHs whose effective temperatures have climbed above the rest energy of hadrons probably give off more photons by *indirect* processes than by direct emission. Rather than emitting bound states (i.e. hadrons), these holes are hot enough to emit elementary constituents (quarks and gluons in the form of relativistic jets). Accelerator experiments and numerical simulations indicate that such jets subsequently fragment into secondary particles whose decays (especially those of the pions) produce a far greater flux of photons than that emitted directly from the PBH. The net effect is to increase the PBH luminosity, particularly in low-energy gamma-rays, strengthening the constraint on $\Omega_{pbh}$ by about an order of magnitude [461]. The most recent upper limit obtained in this way using EGRET data (assuming $\Omega_{m,0} = 1$) is $\Omega_{pbh} < (5.1 \pm 1.3) \times 10^{-9} h_0^{-2}$ [465].

Complementary upper limits on PBH contributions to the dark matter have come from *direct* searches for those evaporating within a few kpc of the Earth. Such limits are subject to more uncertainty than ones based on the EBL because they depend on assumptions about the degree to which PBHs are clustered. If there is no clustering then (9.39) can be converted into a stringent upper bound on the local PBH evaporation rate, $\dot{\mathcal{N}} < 10^{-7}$ pc$^{-3}$ yr$^{-1}$. This however relaxes to $\dot{\mathcal{N}} \lesssim 10$ pc$^{-3}$ yr$^{-1}$ if PBHs are strongly clustered [458], in which case limits from direct searches could potentially become competitive with those based on the EBL. Data taken at energies near 50 TeV with the CYGNUS air-shower array has led to a bound of $\dot{\mathcal{N}} < 8.5 \times 10^5$ pc$^{-3}$ yr$^{-1}$ [466], and a comparable limit of $\dot{\mathcal{N}} < (3.0 \pm 1.0) \times 10^6$ pc$^{-3}$ yr$^{-1}$ has been obtained at 400 GeV using an imaging atmospheric Čerenkov technique developed by the Whipple collaboration [467]. Strong constraints have also been claimed based on balloon observations of cosmic-ray antiprotons [468].

Other ideas have been advanced which could weaken the bounds on PBHs as dark-matter candidates. It might be, for instance, that these objects leave behind stable relics rather than evaporating completely [469], a possibility that has recently been revived on the grounds that total evaporation would be inconsistent with a generalized gravitational version of the uncertainty principle [470]. This however raises a new problem (similar to the "gravitino problem" discussed in Sec. 8.5), because such relics would have been overproduced by quantum and thermal fluctuations in the early Universe. Inflation can be invoked to reduce their density, but must be finely tuned if the same relics are to make up a significant fraction of the dark matter today [471].

A different suggestion due to A.F. Heckler [472, 473] is that particles emitted from the black hole might interact strongly enough above a critical temperature to form a photosphere. This would make the PBH appear cooler as seen from a distance than its actual surface temperature, just as the solar photosphere makes the Sun appear cooler than its core. (In the case of the black hole, however, one has not only an electromagnetic photosphere but a QCD "gluosphere.") The reality of this effect is still under debate [465], but preliminary calculations indicate that it could reduce the intensity of PBH contributions to the gamma-ray background by 60% at 100 MeV, and by as much as two orders of magnitude at 1 GeV [474].

Finally, as discussed already in Sec. 9.1, the limits obtained above can be weakened or evaded if PBH formation occurs in such a way as to produce fewer low-mass objects. The challenge faced in such proposals is to explain how a distribution of this kind comes about in a natural way. A common procedure is to turn the question around and use observational data on the present intensity of the gamma-ray background as a probe of the original PBH formation mechanism. Such an approach has been applied, for example, to put constraints on the spectral index of density fluctuations in the context of PBHs which form via critical collapse [462] or inflation with a "blue" or tilted spectrum [475]. Thus, whether or not they are significant dynamically, PBHs serve as a valuable window on conditions in the early Universe, where information is otherwise scarce.

## 9.6 Higher dimensions

In view of the fact that conventional black holes are disfavored as dark-matter candidates, it is worthwhile to consider alternatives. One of the simplest of these is the extension of the black-hole concept from the 4D spacetime of general relativity to higher dimensions. Higher-dimensional relativity, also known as Kaluza-Klein gravity, has a long history and underlies modern attempts to unify gravity with the standard model of particle physics [476]. The extra dimensions have traditionally been assumed to be compact, in order to explain their non-appearance in low-energy physics. However, the past decade has witnessed a surge of interest in *non*-compactified theories of higher-dimensional gravity [477–479]. In such theories the dimensionality of spacetime can manifest itself at experimentally accessible energies. We focus on the prototypical 5D case, although the extension to higher dimensions is straightforward in principle.

Black holes are described in 4D general relativity by the Schwarzschild metric, which reads (in isotropic coordinates)

$$ds^2 = \left(\frac{1 - GM_s/2c^2r}{1 + GM_s/2c^2r}\right)^2 c^2 dt^2 - \left(1 + \frac{GM_s}{2c^2r}\right)^4 (dr^2 + r^2 \, d\Omega^2) , \quad (9.45)$$

where $d\Omega^2 \equiv d\theta^2 + \sin^2\theta d\phi^2$. This is a description of the static, spherically-symmetric spacetime around a pointlike object (such as a collapsed star or primordial density fluctuation) with Schwarzschild mass $M_s$. As we have seen, it is unlikely that such objects can make up the dark matter.

If the Universe has more than four dimensions, then the same object must be modelled with a higher-dimensional analog of the Schwarzschild metric. Various possibilities have been explored over the years, with most attention focusing on a 5D solution first discussed in detail by D.J. Gross and M.J. Perry [480], R.D. Sorkin [481] and A. Davidson and D.A. Owen [482] in the early 1980s. This is now generally known as the *soliton metric* and reads:

$$ds^2 = \left(\frac{ar - 1}{ar + 1}\right)^{2\xi\kappa} c^2 dt^2 - \left(\frac{a^2 r^2 - 1}{a^2 r^2}\right)^2 \left(\frac{ar + 1}{ar - 1}\right)^{2\xi(\kappa-1)} (dr^2 + r^2 \, d\Omega^2)$$

$$- \left(\frac{ar + 1}{ar - 1}\right)^{2\xi} dy^2 . \quad (9.46)$$

Here $y$ is the new coordinate and there are three metric parameters $(a, \xi, \kappa)$ rather than just one $(M_s)$ as in Eq. (9.45). Only two of these are independent, however, because a consistency condition (which follows from the field equations) requires that $\xi^2(\kappa^2 - \kappa + 1) = 1$. In the limit where $\xi \to 0$, $\kappa \to \infty$ and $\xi\kappa \to 1$, Eq. (9.46) reduces to (9.45) on 4D hypersurfaces $y = \text{const}$. In this limit we can also identify the parameter $a$ as $a = 2c^2/GM_s$ where $M_s$ is the Schwarzschild mass.

We wish to understand the physical properties of this solution in four dimensions. To accomplish this we do two things. First, we assume that Einstein's field equations in their usual form hold in the full *five*-dimensional spacetime. Second, we assume that the Universe in five dimensions is *empty*, with no 5D matter fields or cosmological constant. The field equations then simplify to

$$\mathcal{R}_{AB} = 0 . \quad (9.47)$$

Here $\mathcal{R}_{AB}$ is the 5D Ricci tensor, defined in exactly the same way as the 4D one except that spacetime indices $A, B$ run over 0–4 instead of 0–3. Putting a 5D metric such as (9.46) into the vacuum 5D field equations (9.47), we recover the 4D field equations (5.1) with a nonzero energy-momentum tensor $\mathcal{T}_{\mu\nu}$. *Matter and energy, in other words, are induced in 4D by pure geometry in 5D*. It is by studying the properties of this induced-matter energy-momentum tensor $(T_{\mu\nu})$ that we learn what the soliton looks like in four dimensions.

The details of the mechanism just outlined [483] and its application to solitons in particular [484, 485] have been well studied and we do not review that material here. It is important to note, however, that the Kaluza-Klein soliton differs from an ordinary black hole in several key respects. It contains a singularity at its center, but this center is located at $r = 1/a$ rather than $r = 0$. (The point $r = 0$ is, in fact, not even part of the manifold, which ends at $r = 1/a$.) Its event horizon also shrinks to a point at $r = 1/a$. For these reasons the soliton is better classified as a naked singularity than a black hole.

Solitons in the induced-matter picture are further distinguished from conventional black holes by the fact that they have an extended matter distribution rather than having all their mass compressed into the singularity. It is this feature which proves to be of most use to us in putting constraints on solitons as dark-matter candidates [486]. The time-time component of the induced-matter energy-momentum tensor gives us the density of the solitonic fluid as a function of radial distance:

$$\rho_s(r) = \frac{c^2 \xi^2 \kappa a^6 r^4}{2\pi G (ar - 1)^4 (ar + 1)^4} \left( \frac{ar - 1}{ar + 1} \right)^{2\xi(\kappa - 1)} . \tag{9.48}$$

From the other elements of $T_{\mu\nu}$ one finds that pressure can be written $p_s = \frac{1}{3}\rho_s c^2$, so that the soliton has a radiation-like equation of state. In this respect the soliton more closely resembles a primordial black hole (which forms during the radiation-dominated era) than one which arises as the endpoint of stellar collapse. The elements of $T_{\mu\nu}$ can also be used to calculate the gravitational mass of the fluid inside $r$:

$$M_g(r) = \frac{2c^2 \xi \kappa}{Ga} \left( \frac{ar - 1}{ar + 1} \right)^{\xi} . \tag{9.49}$$

At large distances $r \gg 1/a$ from the center, the soliton's density (9.48) and gravitational mass (9.49) go over to

$$\rho_s(r) \to \frac{c^2 \xi^2 \kappa}{2\pi G a^2 r^4} \qquad M_g(r) \to M_g(\infty) = \frac{2c^2 \xi \kappa}{Ga} . \tag{9.50}$$

The second of these expressions shows that the asymptotic value of $M_g$ is in general not the same as $M_s$ $[M_g(\infty) = \xi\kappa M_s$ for $r \gg 1/a]$, but reduces to it in the limit $\xi\kappa \to 1$. Viewed in four dimensions, the soliton resembles a hole in the geometry surrounded by a spherically-symmetric ball of ultra-relativistic matter whose density falls off at large distances as $1/r^4$. If the

Universe does have more than four dimensions, then objects like this should be common, being generic to 5D Kaluza-Klein gravity in exactly the same way that black holes are to 4D general relativity.

We therefore assess their impact on the background radiation, assuming that the fluid making up the soliton is in fact composed of photons (although one might also consider ultra-relativistic particles such as neutrinos in principle). We do not have spectral information on these so we proceed bolometrically. Putting the second of Eqs. (9.50) into the first gives

$$\rho_s(r) \approx \frac{GM_g^2}{8\pi c^2 \kappa \, r^4} \, . \tag{9.51}$$

Numbers can be attached to the quantities $\kappa, r$ and $M_g$ as follows. The first $(\kappa)$ is technically a free parameter. However, a natural choice from the physical point of view is $\kappa \sim 1$. For this case the consistency relation implies $\xi \sim 1$ also, guaranteeing that the asymptotic gravitational mass of the soliton is close to its Schwarzschild one. To obtain a value for $r$, let us assume that solitons are distributed homogeneously through space with average separation $d$ and mean density $\bar{\rho}_s = \Omega_s \rho_{\text{crit},0} = M_s/d^3$. Since $\rho_s$ drops as $r^{-4}$ whereas the number of solitons at a distance $r$ climbs only as $r^3$, the local density of solitons is largely determined by the nearest one. We can therefore replace $r$ by $d = (M_s/\Omega_s \rho_{\text{crit},0})^{1/3}$. The last unknown in (9.51) is the soliton mass $M_g$ $(= M_s$ if $\kappa = 1)$. The fact that $\rho_s \propto r^{-4}$ is reminiscent of the density profile of the Galactic dark-matter halo, Eq. (6.12). Theoretical work on the classical tests of 5D general relativity [487] and limits on violations of the equivalence principle [488] also suggests that solitons are likely to be associated with dark matter on galactic or larger scales. Let us therefore express $M_s$ in units of the mass of the Galaxy, which from (6.13) is $M_{\text{gal}} \approx 2 \times 10^{12} M_\odot$. Eq. (9.51) then gives the local energy density of solitonic fluid as

$$\rho_s c^2 \approx (3 \times 10^{-17} \text{ erg cm}^{-3}) \, h_0^{8/3} \, \Omega_s^{4/3} \left( \frac{M_s}{M_{\text{gal}}} \right)^{2/3} \, . \tag{9.52}$$

To get a characteristic value, we take $M_s = M_{\text{gal}}$ and adopt our usual values $h_0 = 0.75$ and $\Omega_s = \Omega_{\text{cdm}} = 0.3$. Let us moreover compare our result to the average energy density of the CMB, which dominates the spectrum of background radiation (Fig. 1.15). The latter is found from Eq. (5.37) as $\rho_{\text{cmb}} c^2 = \Omega_\gamma \rho_{\text{crit},0} c^2 = 4 \times 10^{-13} \text{ erg cm}^{-3}$. We therefore obtain

$$\rho_s/\rho_{\text{cmb}} \approx 7 \times 10^{-6} \, . \tag{9.53}$$

This is of the same order of magnitude as the limit set on anomalous contributions to the CMB by COBE and other experiments. Thus the dark matter could consist of solitons, if they are not more massive than galaxies. Similar arguments can be made on the basis of tidal effects and gravitational lensing [486]. To go further, and put more detailed constraints on these candidates from background radiation (or other considerations) will require a deeper investigation of their microphysical properties.

Let us summarize our results for this chapter. We have noted that standard (stellar) black holes cannot provide the dark matter, insofar as their contributions to the density of the Universe are effectively baryonic. Primordial black holes evade this constraint, but we have reconfirmed the classic results of Page, Hawking and others: the collective density of such objects must be negligible, for otherwise their presence would have been obvious in the gamma-ray background. In fact, we have shown that their bolometric intensity alone is sufficient to rule them out as important dark-matter candidates. These constraints may be relaxed if primordial black holes can form in such a way that they are distributed with larger masses, but it is not clear that such a distribution can be shown to arise in a natural way. As an alternative, we have considered objects *like* black holes in higher-dimensional gravity. Bolometric arguments do not rule these out, but there are a number of theoretical issues to be clarified before a more definitive assessment of them can be made.

Chapter 10

# The Universe Seen Darkly

Thus do we arrive—after a series of sometimes technical excursions—back at our starting point. We are now equipped to understand, not only why the night sky is dark, but how its exact level of darkness serves as a kind of "diagnostic tool" for astronomers to peer behind the veil of visible matter and inquire what the Universe is really made of. The intensity of cosmic background radiation contains a wealth of information about the Universe and its contents, both seen and unseen. At near-optical wavelengths, most of this radiation must come from stars in galaxies, for integrating over the known galaxy population (and allowing for galaxy evolution), we obtain theoretical levels of extragalactic background light that are compatible with observations. For *any* realistic combination of parameters, the intensity of this background light is determined to order of magnitude by the age of the Universe, which limits the amount of light that galaxies have been able to produce. Expansion reduces this intensity, but only by a factor of about two, depending on the details of galaxy evolution and the ratio of dark matter to dark energy in the Universe.

The new era of "precision cosmology" has brought us closer to knowing just what this ratio is. The Universe appears to consist of roughly three parts vacuum-like dark energy and one part pressureless cold dark matter, with a sprinkling of hot dark matter (neutrinos) that is almost certainly much less important than cold dark matter. Baryons—the stuff of which we are made—turn out to be mere trace elements in comparison. This marks a fundamental shift in cosmological thinking: our composition is special, even if our location in space is not. The observations do not tell us what dark energy or dark matter are made of, nor *why* these ingredients exist in the ratios they do, a question that is particularly nagging since their densities evolve so differently with time. At present it simply seems that we have stumbled onto the cosmic stage at an unusual moment.

At wavelengths other than the optical, the spectrum of background radiation holds equally valuable clues about the Universe's dark side. Leading candidates for dark matter and dark energy are unstable to radiative decay, or interact with photons in other ways that give rise to characteristic signatures in the cosmic background radiation at various wavelengths. Experimental data on the intensity of this background therefore tell us what the unseen Universe can (or cannot) be made of. Dark energy cannot exist in the form of an unstable vacuum that decays into photons, because this would lead to levels of microwave background radiation in excess of those observed. Dark matter cannot consist of axions or neutrinos with rest energies in the eV-range, because these would produce too much infrared, optical or ultraviolet background light, depending on their lifetimes and coupling parameters. It *could* consist of supersymmetric weakly interacting massive particles (WIMPs) such as neutralinos, but data on the x-ray and gamma-ray backgrounds imply that these must be very nearly stable. The same data exclude a significant role for primordial black holes, whose Hawking evaporation produces too much energy at gamma-ray wavelengths. Higher-dimensional analogs of black holes known as solitons are more difficult to constrain, but an analysis based on the integrated intensity of the background radiation at all wavelengths suggests that they could be dark-matter objects if their masses are not larger than those of galaxies.

While these are the leading candidates, the same methods can be applied to many others as well. We mention some of these here without going into details. Some of the baryonic dark matter could be bound up in an early generation of stars with masses in the $100 - 10^5 M_\odot$ range. These objects, sometimes termed *very massive objects* or VMOs [489, 490], are primarily constrained by their contributions to the infrared background [491]. *Warm dark-matter* (WDM) particles in the keV rest-energy range, including certain types of gravitinos and "sterile" neutrinos, would help to resolve problems in structure formation and might be detected in the x-ray background [492, 493]. Decaying dark-matter particles could have partially reionized the Universe at high redshifts, helping to explain the surprisingly large optical depth inferred from some analyses of CMB data [494–498]. In a similar vein, the recent detection of 511 keV gamma-rays from the Galactic bulge by the INTEGRAL satellite [499] has been interpreted as evidence for a population of annihilating [500] or decaying [501] dark-matter particles with rest energies of $1 - 100$ MeV.

Weakly interacting massive particles do not only arise in supersymmetric theories. Other species of WIMPs that have been proposed include charged massive particles or CHAMPs [502] and technibaryons [503], which are constrained by data on the x-ray and gamma-ray backgrounds respectively. There are particles like "WIMPzillas" [504], very massive, non-thermal WIMPs with rest energies in the $10^{12} - 10^{16}$ GeV range. Their decays would be seen at the upper reaches of the diffuse gamma-ray spectrum [505], and might be responsible for otherwise puzzling observations of ultrahigh-energy cosmic rays [506]. Variations on this theme include: strongly-interacting WIMPzillas or "SIMPzillas" [507], gluinos [508] and axinos [509] (the supersymmetric counterparts of gluons and axions), leptonic WIMPs or "LIMPs" [510], "superWIMPs" [511] (superweakly interacting WIMPs whose existence would only be betrayed by the decays of their parent particles, the *next*-to-lightest SUSY particles), and electromagnetically-coupled or "EWIMPs" [512]. All of these particles would affect primarily the gamma-ray portion of the EBL spectrum.

High-energy gamma-rays also provide the best hunting-ground for the dark-matter candidates that arise generically in higher-dimensional theories [476]. In brane-world models [513], where gravity propagates in a higher-dimensional bulk while all other fields are restricted to the four-dimensional brane we call spacetime, the graviton possesses a tower of massive Kaluza-Klein excitations or *Kaluza-Klein gravitons*. These can carry energy out of supernovae cores, before eventually decaying into photon pairs and other particles. Radiative decays are particularly conspicuous near 30 MeV, and the observed EBL intensity at this energy currently sets the strongest experimental constraints on brane-world scenarios with two and three extra dimensions [514, 515]. Massive brane fluctuations or "branons" are also dark-matter candidates, whose annihilations would show up in the gamma-ray background [516]. In "universal extra dimensions" (UED) models, where all fields can propagate in the bulk, the *lightest Kaluza-Klein particle* or LKP (no longer necessarily related to the graviton) becomes a dark-matter candidate [517]. Such particles have a long history, and were originally known as "pyrgons" [518]. They too turn out to be sharply constrained by their annihilations into gamma-rays [519–521]. Higher-dimensional string and M-theories imply the existence of other superheavy metastable states (with such names as "cryptons," "hexons," "pentons" and "tetrons"). These could also be the dark matter, as well as being responsible for ultrahigh-energy cosmic rays [522–525].

Flights of fancy as some of the above possibilities may be, we mention them to underscore the fact that the study of the dark night sky is open-ended. This in the sense that any suggested new component of the Universe will have to be checked against the fundamental fact that intergalactic space is close to black. The "zeroth-order" reason for this, based on the contents of the preceding chapters, is that space is illuminated primarily by the photons from stars in galaxies, so any other plausible source has to "squeeze in" under the bar set by observations at various wavelengths. The extragalactic background "light" is thus a basic discriminator for variations on standard cosmology.

This is in accordance with that old standby of philosophy, namely Occam's razor, which effectively keeps physics on a productive path by excluding hypotheses which are excessively speculative. At present, this principle is usually applied in the rather crude sense of carrying out experiments to rule out a theoretical possibility. However, not all experiments and observations are clear-cut, and their interpretations are often subject to doubt. An equally valid application of Occam's razor is in the domain of ideas, in the sense that a new suggestion which runs counter to an established body of theory is unlikely to be valid. The study of the dark night sky—both historically and recently—shows many cases where a new hypothesis comes from a narrow base but runs into a broad wall of established physics. At the risk of sounding conservative, not all ideas are "good" ideas.

That said, we cannot deny that our understanding of the darkness of the night sky is due as much to human thought as it is to experiment. We saw in Chap. 1 how that great dreamer Kepler first glimpsed the outlines of the problem in 1610 when he was confronted with observations of new stars that challenged his belief in a finite Universe. Stukeley, a humble physician, stumbled upon it anew in 1721 when he began to suspect that all was not well with the static and infinite Universe of his friend Newton. From Stukeley it passed to Halley (who grasped the problem clearly but fumbled with its solution), and from Halley into the public record. De Chéseaux understood it too, and formulated a solution in 1744 that, while incorrect, satisfied astronomers for over a century. It is an accident of history that the problem became known as Olbers' paradox, though his thoughts of 1823 also go to the point. But it was the poet and writer Poe in 1848, and the astronomer Mädler in 1858, who first connected the fact of the dark night sky with the idea of the *birth of the Universe in time*, and thereby laid down one of the cornerstones of modern cosmology.

On a more general level, the solution of problems in physics by "mental arithmetic" was advocated by A.S. Eddington and his followers in the 1930s [526, 527]. He admitted the existence of an objective, external world; but believed that its interpretation depended on subjective, internal factors connected with our human perceptions. For example, the biological fact of having two eyes set apart gives us the ability to judge distance by a close-up analog of what is termed parallax in astronomy. More universally, the development of physics is akin to the activity of a fisherman, in the sense that we only recover from the sea of knowledge those "discoveries" which are larger than the mesh-size of our mental net. The many twists and turns on the road to our understanding of the dark night sky bear out the essential truth of Eddington's analogy. However, we do not wish here to take sides on the issue of whether our view of the Universe owes more to experiment or thought. We are content with the conclusion at which we arrive by a combination of both: the night sky in a general way is "dark" because the Universe is young.

# Bibliography

[1] H. Bondi, *Cosmology* (Cambridge: Cambridge University Press, 1952)

[2] H.W.M. Olbers, *Astronomisches Jahrbuch für das Jahr 1826* (Berlin: C.F.E. Späthen, 1823)

[3] O. Struve, *Sky and Telescope* **25**, 140 (1963)

[4] S.L. Jaki, *The Paradox of Olbers' Paradox* (New York: Herder, 1969; 2d. ed. Pinckney, MI: Real View Books, 2001)

[5] E.R. Harrison, *Darkness at Night* (Cambridge, MA: Harvard University Press, 1987)

[6] S.L. Jaki, *Am. J. Phys.* **35**, 200 (1967)

[7] S.J. Jaki, *J. Hist. Astron.* **1**, 53 (1970)

[8] M. Hoskin, *J. Br. Astron. Assoc.* **83**, 254 (1973)

[9] D.D. Clayton, *The Dark Night Sky* (New York: Demeter Press, 1975)

[10] E.R. Harrison, *Am. J. Phys.* **45**, 119 (1977)

[11] E.R. Harrison, *Science* **226**, 941 (1984)

[12] M. Hoskin, *J. Hist. Astron.* **16**, 77 (1985)

[13] P.S. Wesson, *Sp. Sci. Rev.* **44**, 169 (1986)

[14] F.J. Tipler, *Q. J. R. Astron. Soc.* **29** 313 (1988)

[15] S.L. Jaki, *Olbers Studies* (Tucson, AZ: Pachart Publishing House, 1991)

[16] M. Hoskin (ed.), *Cambridge Illustrated History of Astronomy* (Cambridge: Cambridge University Press, 1997) 220

[17] M. Hoskin, in V.J. Martínez et al.(eds), *Historical Development of Modern Cosmology* (ASP Conference Series, V. 252, 2001) 11

[18] H. Kragh, *Conceptions of Cosmos* (Oxford: Oxford University Press, 2007)

[19] D.W. Sciama, *The Unity of the Universe* (London: Faber and Faber, 1959)

[20] E. Halley, *Phil. Trans. R. Soc. Lond.* **31**, 22 (dated 1720 but actually read on 9 March 1721)

[21] J.P. Loys de Chéseaux, in *Traité de la comète qui a paru en Décembre 1743 & en Janvier, Fevrier & Mars 1744* (Geneva: Marc-Michel Bousquet & Compagnie, 1744)

[22] J.H. Mädler, *Der Fixsternhimmel* (Leipzig: F.A. Brockhaus, 1858)

[23] W. Thomson (Lord Kelvin), *Philosophical Magazine, Ser. 6* **2**, 161 (1901)

[24] E.R. Harrison, *Phys. Today* **28**, 69 (1975)

[25] M. Born, *Einstein's Theory of Relativity* (Berlin: Springer, 1920); Eng. trans. H.L. Brose (New York: E.P. Dutton, 1924), pp. 287-288; note that this suggestion does not appear in the revised English edition (Dover, 1962)

[26] E.R. Harrison, *Nature* **204**, 271 (1964)

[27] E.R. Harrison, *Mon. Not. R. Astron. Soc.* **131**, 1 (1965)

[28] E.R. Harrison, *Phys. Today* **27**, 30 (1974)

[29] H. Bondi, *The Advancement of Science* **12**, 33 (1955)

[30] S. Weinberg, *Gravitation and Cosmology* (New York: Wiley, 1972)

[31] J.A. Peacock, *Cosmological Physics* (Cambridge: Cambridge University Press, 1999)

[32] D. Layzer, *Nature* **209**, 1340 (1966)

[33] A.H. Lategan, *Am. J. Phys.* **46**, 947 (1978)

[34] E.R. Harrison, *Am. J. Phys.* **46**, 948 (1978)

[35] J.E. Felten, *Bull. Am. Astron. Soc.* **12**, 654 (1979)

[36] P.S. Wesson, K. Valle and R. Stabell, *Astrophys. J.* **317**, 601 (1987)

[37] P.S. Wesson, *Astrophys. J.* **367**, 399 (1991)

[38] J. Maddox, *Nature* **349**, 363 (1991)

[39] J.M. Overduin and P.S. Wesson, *Dark Sky, Dark Matter* (Bristol: Institute of Physics, 2003)

[40] M. Fukugita, C.J. Hogan and P.J.E. Peebles, *Astrophys. J.* **503**, 518 (1998)

[41] P. Norberg *et al.*, *Mon. Not. R. Astron. Soc.* **336**, 907 (2002)

[42] M.R. Blanton *et al.*, *Astrophys. J.* **592**, 819 (2003)

[43] J.E. Felten, *Astrophys. J.* **144**, 241 (1966)

[44] R.H. Dicke, *Gravitation and the Universe* (Philadelphia: American Philosophical Society, 1970)

[45] G.F.R. Ellis, *Class. Quant. Grav.* **5**, 891 (1988)

[46] R.J. Adler and J.M. Overduin, *Gen. Rel. Grav.* **37**, 1491 (2005)

[47] M. White and D. Scott, *Astrophys. J.* **459**, 415 (1996)

[48] W. Rindler, *Essential Relativity* (Berlin: Springer-Verlag, 1977)

[49] T. Padmanabhan, *Structure Formation in the Universe* (Cambridge: Cambridge University Press, 1993)

[50] G.C. McVittie and S.P. Wyatt, *Astrophys. J.* **130**, 1 (1959)

[51] G.J. Whitrow and B.D. Yallop, *Mon. Not. R. Astron. Soc.* **127**, 130 (1964)

[52] G.J. Whitrow and B.D. Yallop, *Mon. Not. R. Astron. Soc.* **130**, 31 (1965)

[53] M. Fukugita, C.J. Hogan and P.J.E Peebles, *Nature* **381**, 489 (1996)

[54] M.J. Sawicki, H. Lin and H.K.C. Yee, *Astron. J.* **113**, 1 (1997)

[55] G.N. Toller, *Astrophys. J.* **266**, L79 (1983)

[56] R.C. Henry, *Astrophys. J.* **516**, L49 (1999)

[57] C.F. Lillie and A.N. Witt, *Astrophys. J.* **208**, 64 (1976)

[58] H. Spinrad and R.P.S. Stone, *Astrophys. J.* **226**, 609 (1978)

[59] R.R. Dube, W.C. Wickes and D.T. Wilkinson, *Astrophys. J.* **232**, 333 (1979)

[60] S.P. Boughn and J.R. Kuhn, *Astrophys. J.* **309**, 33 (1986)

[61] P. Jakobsen *et al.*, *Astron. Astrophys.* **139**, 481 (1984)

[62] P.D. Tennyson *et al.*, *Astrophys. J.* **330**, 435 (1988)

[63] J. Murthy *et al.*, *Astron. Astrophys.* **231**, 187 (1990)

[64] M.G. Hauser *et al.*, *Astrophys. J.* **508**, 25 (1998)

[65] E.L. Wright and E.D. Reese, *Astrophys. J.* **545**, 43 (2000)

[66] L. Cambrésy *et al.*, *Astrophys. J.* **555**, 563 (2001)

[67] R.A. Bernstein, W.L. Freedman and B.F. Madore, *Astrophys. J.* **571**, 56 (2002)

[68] K. Mattila, *Astrophys. J.* **591**, 119 (2003)

[69] R. Ellis *et al.*, *Astrophys. J.* **560**, L119 (2001)

[70] R.B. Partridge and P.J.E. Peebles, *Astrophys. J.* **148**, 377 (1967)

[71] N.M. Tinsley, *Astron. Astrophys.* **24**, 89 (1973)

[72] A.G. Bruzual, PhD thesis, University of California, Berkeley (1981)

[73] A.D. Code and G.A. Welch, *Astrophys. J.* **256**, 1 (1982)

[74] Y. Yoshii and F. Takahara, *Astrophys. J.* **326**, 1 (1988)

[75] R.G. Abraham and S. van den Bergh, *Science* **293**, 1273 (2001)

[76] A.L. Kinney *et al.*, *Astrophys. J.* **467**, 38 (1996)

[77] J.E.G. Devriendt, B. Guiderdoni and R. Sadat, *Astron. Astrophys.* **350**, 381 (1999)

[78] C. Pearson and M. Rowan-Robinson, *Mon. Not. R. Astron. Soc.* **283**, 174 (1996)

[79] T. Totani and Y. Yoshii, *Astrophys. J.* **540**, 81 (2000)

[80] T. Totani and Y. Yoshii, *Astrophys. J.* **501**, L177 (1998)

[81] T. Totani *et al.*, *Astrophys. J.* **550**, L137 (2001)

[82] R. Jimenez and A. Kashlinsky, *Astrophys. J.* **511**, 16 (1999)

[83] R.A. Bernstein, W.L. Freedman and B.F. Madore, *Astrophys. J.* **571**, 107 (2002)

[84] J.C. Kapteyn, *Astrophys. J.* **55**, 302 (1922)

[85] J.H. Oort, *Bull. Astron. Inst. Neth.* **6**, 249 (1932)

[86] F. Zwicky, *Astrophys. J.* **86**, 217 (1937)

[87] J.M. Overduin and W. Priester, *Naturwissenschaften* **88**, 229 (2001)

[88] W.L. Freedman *et al.*, *Astrophys. J.* **553**, 47 (2001)

[89] J.A. Willick and P. Batra, *Astrophys. J.* **548**, 564 (2001)

[90] J.R. Herrnstein *et al.*, *Nature* **400**, 539 (1999)

[91] E. Maoz *et al.*, *Nature* **401**, 351 (1999)

[92] K.Z. Stanek, D. Zaritsky and J. Harris, *Astrophys. J.* **500**, L141 (1998)

[93] E.L. Fitzpatrick *et al.*, *Astrophys. J.* **587**, 685 (2003)

[94] S. Wanajo *et al.*, *Astrophys. J.* **577**, 853 (2002)

[95] H. Schatz *et al.*, *Astrophys. J.* **579**, 626 (2002)

[96] L.M. Griffiths, A. Melchiorri and J. Silk, *Astrophys. J.* **553**, L5 (2001)

[97] T.K.Suzuki, Y. Yoshii and T.C. Beers, *Astrophys. J.* **540** 99 (2000)

[98] D. Tytler *et al.*, *Physica Scripta* **85**, 12 (2000)

[99] T.M. Tripp, B.D. Savage and E.B. Jenkins, *Astrophys. J.* **534**, L1 (2000)

[100] J. Miralda-Escudé *et al.*, *Astrophys. J.* **471**, 582 (1996)

[101] B.J. Carr and M. Sakellariadou, *Astrophys. J.* **516**, 195 (1998)

[102] P. Schneider, *Astron. Astrophys.* **279**, 1 (1993)

[103] C. Alcock *et al.*, *Astrophys. J.* **542**, 281 (2000)

[104] C. Afonso *et al.*, *Astron. Astrophys.* **400**, 951 (2003)

[105] B.D. Fields, K. Freese and D.S. Graff, *Astrophys. J.* **534**, 265 (2000)

[106] C. Flynn, A. Gould and J.N. Bahcall, *Astrophys. J.* **466**, L55 (1996)

[107] B.M.S. Hansen, *Astrophys. J.* **517**, L39 (1999)

[108] D. Lynden-Bell and C.A. Tout, *Astrophys. J.* **558** 1 (2001)

[109] D.J. Hegyi and K.A. Olive, *Astrophys. J.* **303**, 56 (1986)

[110] C.S. Kochanek, *Astrophys. J.* **457**, 228 (1996)

[111] R.G. Carlberg, H.K.C. Yee and E. Ellingson, *Astrophys. J.* **478**, 462 (1997)

[112] R.G. Carlberg *et al.*, *Astrophys. J.* **516**, 552 (1999)

[113] N.A. Bahcall *et al.*, *Astrophys. J.* **541**, 1 (2000)

[114] N.A. Bahcall *et al.*, *Science* **284**, 1481 (1999)

[115] J.P. Henry, *Astrophys. J.* **534**, 565 (2000)

[116] H. Hoekstra, H.K.C. Yee and M.D. Gladders, *Astrophys. J* **577**, 595 (2002)

[117] J.P. Ostriker and P. Steinhardt, *Science* **300**, 1909 (2003)

[118] N.A. Bahcall and X. Fan, *Astrophys. J.* **504**, 1 (1998)

[119] D.H. Weinberg *et al.*, *Astrophys. J.* **522**, 563 (1999)

[120] J.A. Peacock *et al.*, in J.C. Wheeler and H. Martel (eds.), *Proc. 20th Texas Symposium on Relativistic Astrophysics* (New York: American Institute of Physics, 2001) p. 938

[121] N.A. Bahcall *et al.*, *Astrophys. J.* **585**, 182 (2003)

[122] M. Tegmark *et al.*, *Phys. Rev.* **D69**, 103501 (2004)

[123] S.L. Bridle *et al.*, *Science* **299**, 1532 (2003)

[124] I. Zehavi and A. Dekel, *Nature* **401**, 252 (1999)

[125] T. Padmanabhan, *Theoretical Astrophysics Volume I* (Cambridge: Cambridge University Press, 2000)

[126] D.W. Sciama, *Modern Cosmology and the Dark Matter Problem* (Cambridge: Cambridge University Press, 1993)

[127] E. Gawiser and J. Silk, *Science* **280**, 1405 (1998)

[128] Ø. Elgarøy *et al.*, *Phys. Rev. Lett.* **89**, 061301 (2002)

[129] E.W. Kolb and M.S. Turner, *The Early Universe* (Reading: Addison-Wesley, 1990)

[130] C. Athanassopoulos *et al.*, *Phys. Rev. Lett.* **81**, 1774 (1998)

[131] Y. Fukuda *et al.*, *Phys. Rev. Lett.* **81**, 1562 (1998)

[132] N. Abdurashitov *et al.*, *Phys. Rev. Lett.* **83**, 4686 (1999)

[133] B.T. Cleveland *et al.*, *Astrophys. J.* **496**, 505 (1998)

[134] W. Hampel *et al.*, *Phys. Lett.* **B447**, 127 (1999)

[135] Q.R. Ahmad *et al.*, *Phys. Rev. Lett.* **89**, 011302 (2002)

[136] K. Eguchi *et al.*, *Phys. Rev. Lett.* **90**, 021802 (2003)

[137] M.H. Ahn *et al.*, *Phys. Rev. Lett.* **90**, 041801 (2003)

[138] M. Maltoni *et al.*, *New J. Phys.* **6**, 122 (2004)

[139] H.V. Klapdor-Kleingrothaus *et al.*, *Mod. Phys. Lett.* **A16**, 2409 (2001)

[140] W. Priester, J. Hoell and H.-J. Blome, *Comments Astrophys.* **17**, 327 (1995)

[141] G. Gamow, *My World Line* (New York: Viking Press, 1970)

[142] S.M. Carroll, *Living Rev. Rel.* **4**, http://www.livingreviews.org/Articles/Volume4/2001-1carroll (2001)

[143] H.E. Puthoff, *Phys. Rev.* **A40**, 4857 (1989)

[144] P.S. Wesson, *Astrophys. J.* **378**, 466 (1991)

[145] J.M. Overduin and F.I. Cooperstock, *Phys. Rev.* **D58**, 043506 (1998)

[146] P.S. Wesson and H. Liu, *Int. J. Mod. Phys.* **D10**, 905 (2001)

[147] S. Weinberg, in D.B. Cline (ed.), *Sources and Detection of Dark Matter in the Universe* (Berlin: Springer-Verlag, 2001) p. 18

[148] J.R. Gott III, M.-G. Park and H.M. Lee, *Astrophys. J.* **338**, 1 (1989)

[149] J.-P. Kneib *et al.*, *Astrophys. J.* **607**, 697 (2004)

[150] C.S. Kochanek, *Astrophys. J.* **466**, 638 (1996)

[151] S. Malhotra, J.E. Rhoads and E.L. Turner, *Mon. Not. R. Astron. Soc.* **288**, 138 (1997)

[152] E.E. Falco, C.S. Kochanek and J.A. Muñoz, *Astrophys. J.* **494**, 47 (1998)

[153] C.R. Keeton, *Astrophys. J.* **595**, L1 (2002)

[154] J.L. Mitchell *et al.*, *Astrophys. J.*, submitted (2004)

[155] M. Fukugita *et al.*, *Astrophys. J.* **361**, L1 (1990)

[156] J.P. Gardner, L.L. Cowie and R.J. Wainscoat, *Astrophys. J.* **415**, L9 (1993)

[157] T. Totani, Y. Yoshii and K. Sato, *Astrophys. J.* **483**, L75 (1997)

[158] A. Jenkins *et al.*, *Astrophys. J.* **499**, 20 (1998)

[159] H.A. Feldman and A.E. Evrard, *Int. J. Mod. Phys.* **D2**, 113 (1993)

[160] A.G. Riess *et al.*, *Astron. J.* **116**, 1009 (1998)

[161] S. Perlmutter *et al.*, *Astrophys. J.* **517**, 565 (1999)

[162] J.L. Tonry *et al.*, *Astrophys. J.* **594**, 1 (2003)

[163] R.A. Knop *et al.*, *Astrophys. J.* **598**, 102 (2003)

[164] A.N. Aguirre, *Astrophys. J.* **512**, L19 (1999)

[165] P.S. Drell, T.J. Loredo and I. Wasserman, *Astrophys. J.* **530**, 593 (2000)

[166] A.G. Riess *et al.*, *Astrophys. J.* **607**, 665 (2004)

[167] C.H. Lineweaver, in T.G. Brainerd and C.S. Kochanek (eds.), *Gravitational Lensing: Recent Progress and Future Goals* (San Francisco: Astron. Soc. Pac. Conf. Ser. 237, 2001)

[168] P. de Bernardis *et al.*, *Nature* **404**, 955 (2000)

[169] S. Hanany *et al.*, *Astrophys. J.* **545**, L5 (2000)

[170] C. Pryke *et al.*, *Astrophys. J.* **568**, 46 (2001)

[171] D.N. Spergel *et al.*, *Astrophys. J. Supp.* **148**, 1 (2003)

[172] M. White, D. Scott and E. Pierpaoli, *Astrophys. J.* **545**, 1 (2000)

[173] S.S. McGaugh, *Astrophys. J.* **541**, L33 (2000)

[174] D.-E. Liebscher, W. Priester and J. Hoell, *Astron. Astrophys.* **261**, 377 (1992)

[175] S.S. McGaugh, *Astrophys. J.* **611**, 26 (2004)

[176] A.H. Jaffe *et al.*, *Phys. Rev. Lett.* **86**, 3475 (2000)

[177] M. Bronstein, *Physikalische Zeitschrift der Sowjetunion* **3**, 73 (1933)

[178] H. Kragh, *Cosmology and Controversy* (Princeton: Princeton University Press, 1996)

[179] A. Einstein, *J. Franklin Inst.* **221**, 349 (1936)

[180] P. Jordan, *Nature* **164**, 637 (1949)

[181] M. Fierz, *Helv. Phys. Acta* **29**, 128 (1956)

[182] C. Brans and R.H. Dicke, *Phys. Rev.* **124**, 925 (1961)

[183] P.G. Bergmann, *Int. J. Theor. Phys.* **1**, 25 (1968)

[184] M. Endō and T. Fukui, *Gen. Rel. Grav.* **8**, 833 (1977)

[185] J.D. Barrow and K.-I. Maeda, *Nucl. Phys.* **B341**, 294 (1990)

[186]  T. Fukui and J.M. Overduin, *Int. J. Mod. Phys.* **D11**, 669 (2002)
[187]  M.S. Madsen, *Class. Quant. Grav.* **5**, 627 (1988)
[188]  K.-I. Maeda, *Phys. Rev.* **D39**, 3159 (1989)
[189]  S.M. Barr, *Phys. Rev.* **D36**, 1691 (1987)
[190]  P.J.E. Peebles and B. Ratra, *Astrophys. J.* **325**, L17 (1988)
[191]  C. Wetterich, *Nucl. Phys.* **B302**, 668 (1988)
[192]  R.R. Caldwell, R. Dave and P.J. Steinhardt, *Phys. Rev. Lett.* **80**, 1582 (1998)
[193]  Y.B. Zeldovich, *Uspekhi Fiz. Nauk* **95**, 209 (1968)
[194]  A.M. Polyakov, *Sov. Phys. Usp.* **25**, 187 (1982)
[195]  S.L. Adler, *Rev. Mod. Phys.* **54**, 729 (1982)
[196]  A.D. Dolgov, in G.W. Gibbons, S.W. Hawking and S.T.C. Siklos (eds.), *The Very Early Universe* (Cambridge: Cambridge University Press, 1983) p. 449
[197]  T. Banks, *Nucl. Phys.* **B249**, 332 (1985)
[198]  R.D. Peccei, J. Solà and C. Wetterich, *Phys. Lett.* **195B**, 183 (1987)
[199]  J.A. Frieman *et al.*, *Phys. Rev. Lett.* **75**, 2077 (1995)
[200]  S.W. Hawking, *Phys. Lett.* **134B**, 403 (1984)
[201]  J.D. Brown and C. Teitelboim, *Phys. Lett.* **195B**, 177 (1987)
[202]  A.D. Dolgov, *Phys. Rev.* **55**, 5881 (1997)
[203]  E. Mottola, *Phys. Rev.* **D31**, 754 (1985)
[204]  N.C. Tsamis and R.P. Woodard, *Phys. Lett.* **B301**, 351 (1993)
[205]  V.F. Mukhanov, L.R.W. Abramo and R.H. Brandenberger, *Phys. Rev. Lett.* **78**, 1624 (1997)
[206]  T. Banks, *Nucl. Phys.* **B309**, 493 (1988)
[207]  S. Coleman, *Nucl. Phys.* **B310**, 643 (1988)
[208]  E.I. Guendelman and A.B. Kaganovich, *Phys. Rev.* **D55**, 5970 (1997)
[209]  S. Weinberg, *Rev. Mod. Phys.* **61**, 1 (1989)
[210]  A.D. Dolgov, in N. Sanchez and H.J. de Vega (eds.), *The 4th Paris Cosmology Colloquium* (Singapore: World Scientific, 1998)
[211]  K. Freese *et al.*, *Nucl. Phys.* **B287**, 797 (1987)
[212]  M. Birkel and S. Sarkar, *Astroparticle Phys.* **6**, 197 (1997)
[213]  T.S. Olson and T.F. Jordan, *Phys. Rev.* **D35**, 3258 (1987)
[214]  J. Matyjasek, *Phys. Rev.* **D51**, 4154 (1995)
[215]  B. Ratra and P.J.E. Peebles, *Phys. Rev.* **D37**, 3406 (1988)
[216]  V. Silveira and I. Waga, *Phys. Rev.* **D50**, 4890 (1994)
[217]  P.T.P. Viana and A.R. Liddle, *Phys. Rev.* **D57**, 674 (1998)
[218]  Y. Yoshii and K. Sato, *Astrophys. J.* **387**, L7 (1992)
[219]  N. Sugiyama and K. Sato, *Astrophys. J.* **387**, 439 (1992)
[220]  V. Silveira and I. Waga, *Phys. Rev.* **D56**, 4625 (1997)
[221]  B. Ratra and A. Quillen, *Mon. Not. R. Astron. Soc.* **259**, 738 (1992)
[222]  L.F.B. Torres and I. Waga, *Mon. Not. R. Astron. Soc.* **279**, 712 (1996)
[223]  S. Podariu and B. Ratra, *Astrophys. J.* **532**, 109 (2000)
[224]  J.M. Overduin, *Astrophys. J.* **517**, L1 (1999)
[225]  D. Pavón, *Phys. Rev.* **D43**, 375 (1991)
[226]  J.A.S. Lima, *Phys. Rev.* **D54**, 2571 (1996)

[227] D.J. Fixsen *et al.*, *Astrophys. J.* **473**, 576 (1996)

[228] W. Hu and J. Silk, *Phys. Rev.* **D48**, 485 (1993)

[229] M.D. Pollock, *Mon. Not. R. Astron. Soc.* **193**, 825 (1980)

[230] J.M. Overduin, P.S. Wesson and S. Bowyer, *Astrophys. J.* **404**, 1 (1993)

[231] C.W. Misner, K.S. Thorne and J.A. Wheeler, *Gravitation* (San Francisco: W.H. Freeman, 1973)

[232] P.J.E. Peebles, *Principles of Physical Cosmology* (Princeton: Princeton University Press, 1993)

[233] D.J. Fixsen *et al.*, *Astrophys. J.* **508**, 123 (1998)

[234] G. Lagache *et al.*, *Astron. Astrophys.* **354**, 247 (2000)

[235] R. Peccei and H. Quinn, *Phys. Rev. Lett.* **38**, 1440 (1977)

[236] S. Weinberg, *Phys. Rev. Lett.* **40**, 223 (1978)

[237] F. Wilczek, *Phys. Rev. Lett.* **40**, 279 (1978)

[238] J. Preskill, M. Wise and F. Wilczek, *Phys. Lett.* **B120**, 127 (1983)

[239] L. Abbott and P. Sikivie, *Phys. Lett.* **B120**, 133 (1983)

[240] M. Dine and W. Fischler, *Phys. Lett.* **B120**, 137 (1983)

[241] R. Davis, *Phys. Lett.* **B180**, 255 (1986)

[242] P. Sikivie, in H.V. Klapdor-Kleingrothaus and I.V. Krivosheina (eds.), *Beyond the Desert 1999* (Oxford: Institute of Physics Press, 2000) p. 547

[243] R.A. Battye and E.P.S. Shellard, in H.V. Klapdor-Kleingrothaus and I.V. Krivosheina (eds.), *Beyond the Desert 1999* (Oxford: Institute of Physics Press), p. 565

[244] H.-T. Janka *et al.*, *Phys. Rev. Lett.* **76**, 2621 (1996)

[245] W. Keil *et al.*, *Phys. Rev.* **D56**, 2419 (1997)

[246] P. Sikivie, *Phys. Lett.* **51**, 1415 (1983)

[247] C. Hagmann *et al.*, *Phys. Rev. Lett.* **80**, 2043 (1998)

[248] S. Moriyama *et al.*, *Phys. Lett.* **B434**, 147 (1998)

[249] F.T. Avignone *et al.*, *Phys. Rev. Lett.* **81**, 5068 (1998)

[250] A. Morales *et al.*, *Astropart. Phys.* **16**, 325 (2002)

[251] C.E. Aalseth *et al.*, *Nucl. Phys. B. Proc. Supp.* **110**, 85 (2002)

[252] C. Csáki, N. Kaloper and J. Terning, *Phys. Rev. Lett.* **88** 161302 (2002)

[253] M.T. Ressell, *Phys. Rev.* **D44**, 3001 (1991)

[254] D.B. Kaplan, *Nucl. Phys.* **B260**, 215 (1985)

[255] M.S. Turner, *Phys. Rev. Lett.* **59**, 2489 (1987)

[256] M.S. Turner, *Phys. Rev. Lett.* **60**, 1797 (1988)

[257] J.E. Kim, *Phys. Rev. Lett.* **43**, 103 (1979)

[258] M.A. Shifman, A.I. Vainshtein and V.I. Zakharov, *Nucl. Phys.* **B166**, 493 (1980)

[259] A.R. Zhitnitsky, *Sov. J. Nucl. Phys.* **31**, 260 (1980)

[260] M. Dine, W. Fischler and M. Srednicki, *Phys. Lett.* **B104**, 199 (1981)

[261] G. Raffelt and A. Weiss, *Phys. Rev.* **D51**, 1495 (1995)

[262] G.G. Raffelt and D.S.P Dearborn, *Phys. Rev.* **D36**, 2211 (1987)

[263] J. Engel, D. Seckel and A.C. Hayes, *Phys. Rev.* **D65**, 960 (1990)

[264] W. Jaffe, *Mon. Not. R. Astron. Soc.* **202**, 995 (1983)

[265] J.M. Overduin and P.S. Wesson, *Astrophys. J.* **414**, 449 (1993)

[266] T.W. Kephart and T.J. Weiler, *Phys. Rev. Lett.* **58**, 171 (1987)

[267] M.A. Bershady, M.T. Ressell and M.S. Turner, *Phys. Rev. Lett.* **66**, 1398 (1991)

[268] J.F. Lodenquai and V.V. Dixit, *Phys. Lett.* **B194**, 350 (1987)

[269] B.D. Blout *et al.*, *Astrophys. J.* **546**, 825 (2001)

[270] R. Cowsik, *Phys. Rev. Lett.* **39**, 784 (1977)

[271] A. de Rujula and S.L. Glashow, *Phys. Rev. Lett.* **45**, 942 (1980)

[272] P.B. Pal and L. Wolfenstein, *Phys. Rev.* **D25**, 766 (1982)

[273] S. Bowyer *et al.*, *Phys. Rev.* **D52**, 3214 (1995)

[274] A.L. Melott and D.W. Sciama, *Phys. Rev. Lett.* **46**, 1369 (1981)

[275] D.W. Sciama and A.L. Melott, *Phys. Rev.* **D25**, 2214 (1982)

[276] A.L. Melott, D.W. McKay and J.P. Ralston, *Astrophys. J.* **324**, L43 (1988)

[277] D.W. Sciama, *Astrophys. J.* **364**, 549 (1990)

[278] D.W. Sciama, *Mon. Not. R. Astron. Soc.* **289**, 945 (1997)

[279] A. Fabian, T. Naylor and D. Sciama, *Mon. Not. R. Astron. Soc.* **249**, 21 (1991)

[280] A.F. Davidsen *et al.*, *Nature* **351**, 128 (1991)

[281] D.W. Sciama, *Pub. Astron. Soc. Pac.* **105**, 102 (1993)

[282] F.W. Stecker, *Phys. Rev. Lett.* **45**, 1460 (1980)

[283] R. Kimble, S. Bowyer and P. Jakobsen, *Phys. Rev. Lett.* **46**, 80 (1981)

[284] D.W. Sciama, in B. Rocca-Volmerange, J.M. Deharveng and J.T.T. Van (eds.), *The Early Observable Universe from Diffuse Backgrounds* (Gif-sur-Yvette: Editions Frontières, 1992), p. 127

[285] J.M. Overduin, P.S. Wesson and S. Bowyer, *Astrophys. J.* **404**, 460 (1993)

[286] S. Dodelson and J.M. Jubas, *Mon. Not. R. Astron. Soc.* **266**, 886 (1994)

[287] J.M. Overduin and P.S. Wesson, *Astrophys. J.* **483**, 77 (1997)

[288] J.M. Overduin, S.S. Seahra, W.W. Duley and P.S. Wesson, *Astron. Astrophys.* **349**, 317 (1999)

[289] S. Bowyer, *Ann. Rev. Astron. Astrophys.* **29**, 59 (1991)

[290] R.C. Henry, *Ann. Rev. Astron. Astrophys.* **29**, 89 (1991)

[291] P. Salucci and D.W. Sciama, *Mon. Not. R. Astron. Soc.* **244**, 9P (1990)

[292] E.I. Gates, G. Gyuk and M.S. Turner, *Astrophys. J.* **449**, L123 (1995)

[293] L. Zuo and E.S. Phinney, *Astrophys. J.* **418**, 28 (1993)

[294] J.P. Ostriker and L.L. Cowie, *Astrophys. J.* **243**, L127 (1981)

[295] E.L. Wright, *Astrophys. J.* **250**, 1 (1981)

[296] J.P. Ostriker and J. Heisler, *Astrophys. J.* **278**, 1 (1984)

[297] E.L. Wright, *Astrophys. J.* **311**, 156 (1986)

[298] E.L. Wright and M.A. Malkan, *Bull. Am. Astron. Soc.* **19**, 699 (1987)

[299] J. Heisler and J.P. Ostriker, *Astrophys. J.* **332**, 543 (1988)

[300] E.L. Wright, *Astrophys. J.* **353**, 413 (1990)

[301] S.M. Fall and Y.C. Pei, *Astrophys. J.* **402**, 479 (1993)

[302] J.S. Mathis, *Ann. Rev. Astron. Astrophys.* **28**, 37 (1990)

[303] B.T. Draine and H.M. Lee, *Astrophys. J.* **285**, 89 (1984)

[304] P.G. Martin and F. Rouleau, in R.F. Malina and S. Bowyer (eds.), *Extreme Ultraviolet Astronomy* (New York: Pergamon, 1991), p. 341

[305] J.-P. Meyer, *Les Elements et leurs Isotopes dan L'univers* (Liège: Université de Liège, 1979), p. 153

[306] T.P. Snow and A.N. Witt, *Astrophys. J.* **468**, L65 (1996)

[307] B.T. Draine, http://www.astro.princeton.edu/~draine/dust/ (1995)

[308] J.S. Mathis, *Astrophys. J.* **472**, 643 (1996)

[309] W.W. Duley and S. Seahra, *Astrophys. J.* **507**, 874 (1998)

[310] W.W. Duley and S.S. Seahra, *Astrophys. J.* **522**, L129 (1999)

[311] S.S. Seahra and W.W. Duley, *Astrophys. J.* **520**, 719 (1999)

[312] C. Martin and S. Bowyer, *Astrophys. J.* **338**, 677 (1989)

[313] T.P. Sasseen, M. Lampton, S. Bowyer and X. Wu, *Astrophys. J.* **447**, 630 (1995)

[314] A.M. Zvereva *et al.*, *Astron. Astrophys.* **116**, 312 (1982)

[315] F. Paresce, B. Margon, S. Bowyer and M. Lampton, *Astrophys. J.* **230**, 304 (1979)

[316] R.C. Anderson *et al.*, *Astrophys. J.* **234**, 415 (1979)

[317] P.D. Feldman, W.H. Brune and R.C. Henry, *Astrophys. J.* **249**, L51 (1991)

[318] C.S. Weller, *Astrophys. J.* **268**, 899 (1983)

[319] C. Martin, M. Hurwitz and S. Bowyer, *Astrophys. J.* **379**, 549 (1991)

[320] R.C. Henry and J. Murthy, *Astrophys. J.* **418**, L17 (1993)

[321] J.D. Fix, J.D. Craven and L.A. Frank, *Astrophys. J.* **345**, 203 (1989)

[322] E.L. Wright, *Astrophys. J.* **391**, 34 (1992)

[323] A.N. Witt and J.K. Petersohn, in R.M. Cutri and W.B. Latter (eds.), *The First Symposium on the Infrared Cirrus and Diffuse Interstellar Clouds* (San Francisco: Astron. Soc. Pac. Conf. Ser. 58, 1994), p. 91

[324] J.B. Holberg, *Astrophys. J.* **311**, 969 (1986)

[325] J. Murthy, R.C. Henry and J.B. Holberg, *Astrophys. J.* **383**, 198 (1991)

[326] J. Murthy *et al.*, *Astrophys. J.* **522**, 904 (1999)

[327] J. Edelstein, S. Bowyer and M. Lampton, *Astrophys. J.* **539**, 187 (2000)

[328] J. Murthy *et al.*, *Astrophys. J.* **557**, L47 (2001)

[329] E.J. Korpela, S. Bowyer and J. Edelstein, *Astrophys. J.* **495**, 317 (1998)

[330] J. Edelstein *et al.*, *Astrophys. Space Sci.* **276**, 177 (2001)

[331] S. Bowyer *et al.*, *Astrophys. J.* **526**, 10 (1999)

[332] G.G. Raffelt, *Phys. Rev. Lett.* **81**, 4020 (1998)

[333] G. Jungman, M. Kamionkowski and K. Griest, *Phys. Rep.* **267**, 195 (1996)

[334] N. Cabibbo, G.R. Farrar and L. Maiani, *Phys. Lett.* **105B**, 155 (1981)

[335] H. Pagels and J.R. Primack, *Phys. Rev. Lett.* **48**, 223 (1982)

[336] S. Weinberg, *Phys. Rev. Lett.* **48**, 1303 (1982)

[337] J. Ellis, A.D. Linde and D.V. Nanopoulos, *Phys. Lett.* **118B**, 59 (1982)

[338] H. Goldberg, *Phys. Rev. Lett.* **50**, 1419 (1983)

[339] L.E. Ibáñez, *Phys. Lett.* **137B**, 160 (1984)

[340] D.W. Sciama, *Phys. Lett.* **137B**, 169 (1984)

[341] J. Ellis *et al.*, *Nucl. Phys.* **B238**, 453 (1984)

[342] G. Steigman and M.S. Turner, *Nucl. Phys.* **B253**, 375 (1985)

[343] J. Ellis and K.A. Olive, *Phys. Lett.* **B514**, 114 (2001)

[344] J. Ellis *et al.*, *Phys. Rev.* **D62**, 075010 (2000)

[345] K. Griest, M. Kamionkowski and M.S. Turner, *Phys. Rev.* **D41**, 3565 (1990)

[346] J. Ellis, T. Falk and K.A. Olive, *Phys. Lett.* **B444**, 367 (1998)

[347] L. Roszkowski, R.R. de Austri and T. Nihei, *J. High Energy Phys.* **0108**, 024 (2001)

[348] R. Bernabei *et al.*, *Phys. Lett.* **B480**, 23 (2000)
[349] A. Benoit *et al.*, *Phys. Lett.* **B513**, 15 (2001)
[350] D.S. Akerib *et al.*, *Phys. Rev.* **D68**, 082002 (2003)
[351] D.S. Akerib *et al.*, *Phys. Rev. Lett.* **93**, 211301 (2004)
[352] J.C. Barton *et al.*, in N.J.C. Spooner and V. Kudryavtsev (eds.), *The Identification of Dark Matter* (Singapore: World Scientific, 2003), p. 302
[353] I.G. Irastorza *et al.*, in N.J.C. Spooner and V. Kudryavtsev (eds.), *The Identification of Dark Matter* (Singapore: World Scientific, 2003), p. 308
[354] J. Silk and M. Srednicki, *Phys. Rev. Lett.* **53**, 624 (1984)
[355] J. Silk, K. Olive and M. Srednicki, *Phys. Rev. Lett.* **55**, 257 (1985)
[356] K. Freese, *Phys. Lett.* **167B**, 295 (1986)
[357] L.M. Krauss, M. Srednicki and F. Wilczek, *Phys. Rev.* **D33**, 2079 (1986)
[358] J.X. Ahrens *et al.*, *Phys. Rev.* **D66**, 032006 (2002)
[359] J. Silk and H. Bloemen, *Astrophys. J.* **313**, L47 (1987)
[360] F.W. Stecker, *Phys. Lett.* **B201**, 529 (1988)
[361] S. Rudaz, *Phys. Rev.* **D39**, 3549 (1989)
[362] F.W. Stecker and A.J. Tylka, *Astrophys. J.* **343**, 169 (1989)
[363] A. Bouquet, P. Salati and J. Silk, *Phys. Rev.* **D40**, 3168 (1989)
[364] L. Bergström, *Nucl. Phys.* **B325**, 647 (1989)
[365] K. Freese and J. Silk, *Phys. Rev.* **D40**, 3828 (1989)
[366] V. Berezinsky, A. Bottino and G. Mignola, *Phys. Lett.* **B325**, 136 (1994)
[367] P. Gondolo and J. Silk, *Phys. Rev. Lett.* **83**, 1719 (1999)
[368] C. Calcáneo-Roldán and B. Moore, *Phys. Rev.* **D62**, 123005 (2000)
[369] V. Berezinsky, V. Dokuchaev and Y. Eroshenko, *Phys. Rev.* **D68**, 103003 (2003)
[370] S.M. Koushiappas, A.R. Zentner and T.P. Walker, *Phys. Rev.* **D69**, 043501 (2004)
[371] F. Prada *et al.*, *Phys. Rev. Lett.* **93**, 241301 (2004)
[372] N.W. Evans, F. Ferrer and S. Sarkar, *Phys. Rev.* **D69**, 123501 (2004)
[373] F. Stoehr *et al.*, *Mon. Not. R. Astron. Soc.* **345**, 1313 (2003)
[374] M. Urban *et al.*, *Phys. Lett.* **B293**, 149 (1992)
[375] P. Chardonnet *et al.*, *Astrophys. J.* **454**, 774 (1995)
[376] Tsuchiya *et al.*, *Astrophys. J.* **606**, L115 (2004)
[377] K. Kosack *et al.*, *Astrophys. J.* **608**, L97 (2004)
[378] R. Atkins *et al.*, *Phys. Rev.* **D70**, 083516 (2004)
[379] G. Lake, *Nature* **346**, 39 (1990)
[380] P. Gondolo, *Nucl. Phys. Proc. Suppl.* **B35**, 148 (1994)
[381] E.A. Baltz *et al.*, *Phys. Rev.* **D61**, 023514 (1999)
[382] D.B. Cline and Y.-T. Gao, *Astron. Astrophys.* **231**, L23 (1990)
[383] Y.-T. Gao, F.W. Stecker and D.B. Cline, *Astron. Astrophys.* **249**, 1 (1991)
[384] J.M. Overduin and P.S.Wesson, *Astrophys. J.* **480**, 470 (1997)
[385] L. Bergström, P. Ullio and J.H. Buckley, *Astropart. Phys.* **9**, 137 (1998)
[386] V.S. Berezinsky, A. Bottino and V. de Alfaro, *Phys. Lett.* **B274**, 122 (1992)
[387] G.F. Giudice and K. Griest, *Phys. Rev.* **D40**, 2549 (1989)
[388] M.R. Metzger, J.A.R. Caldwell and P.L. Schechter, *Astron. J.* **115**, 635 (1998)

[389] V.S. Berezinsky, A.V. Gurevich and K.P. Zybin, *Phys. Lett.* **B294**, 221 (1992)

[390] A. Bouquet and P. Salati, *Nucl. Phys.* **B284**, 557 (1987)

[391] A. Masiero and J.W.F. Valle, *Phys. Lett.* **B251**, 273 (1990)

[392] V. Berezinsky, A. Masiero and J.W.F. Valle, *Phys. Lett.* **B266**, 382 (1991)

[393] R. Barbieri and V. Berezinsky, *Phys. Lett.* **B205**, 559 (1988)

[394] G.R. Blumenthal and R.J. Gould, *Rev. Mod. Phys.* **42**, 237 (1970)

[395] V.S. Berezinsky, *Nucl. Phys.* **B380**, 478 (1992)

[396] R. Svensson and A.A. Zdziarski, *Astrophys. J.* **349**, 415 (1990)

[397] P.S. Coppi and F.A. Aharonian, *Astrophys. J.* **487**, L9 (1997)

[398] G.D. Kribs and I.Z. Rothstein, *Phys. Rev.* **D55**, 4435 (1997)

[399] R.J. Protheroe, T. Stanev and V.S. Berezinsky, *Phys. Rev.* **D51**, 4134 (1995)

[400] J. Ellis, E. Kim and D.V. Nanopoulos, *Phys. Lett.* **B145**, 181 (1984)

[401] M. Kawasaki and T. Moroi, *Prog. Theor. Phys.* **93**, 879 (1995)

[402] G.F. Giudice, E.M. Kolb and A. Riotto, *Phys. Rev.* **D64**, 023508 (2001)

[403] V.A. Kuzmin, V.A. Rubakov and M.E. Shaposhnikov, *Phys. Lett.* **B155**, 36 (1985)

[404] S. Dimopoulos and L.J. Hall, *Phys. Lett.* **B196**, 135 (1987)

[405] S. Dimopoulos *et al.*, *Astrophys. J.* **330**, 545 (1988)

[406] V.S. Berezinsky, *Phys. Lett.* **B261**, 71 (1991)

[407] M. Markevitch *et al.*, *Astrophys. J.* **583**, 70 (2003)

[408] R. Lieu *et al.*, *Astrophys. J.* **417**, L41 (1993)

[409] D.E. Gruber, in X. Barcon and A.C. Fabian (eds.), *The X-Ray Background* (Cambridge: Cambridge University Press, 1992), p. 44

[410] D.E. Gruber, *Astrophys. J.* **520**, 124 (1999)

[411] D.H. Lumb *et al.*, *Astron. Astrophys.* **389**, 93 (2002)

[412] M. Revnivtsev *et al.*, *Astron. Astrophys.* **411**, 329 (2003)

[413] A.A. Zdziarski, *Mon. Not. R. Astron. Soc.* **281**, L9 (1996)

[414] S.C. Kappadath *et al.*, *Astron. Astrophys. Suppl. Ser.* **120**, 619 (1996)

[415] L.-S. The, M.D. Leising and D.D. Clayton, *Astrophys. J.* **403**, 32 (1993)

[416] D.J. Thompson and C.E. Fichtel, *Astron. Astrophys.* **109**, 352 (1982)

[417] P. Sreekumar *et al.*, *Astrophys. J.* **494**, 523 (1998)

[418] K. McNaron-Brown *et al.*, *Astrophys. J.* **451**, 575 (1995)

[419] U. Keshet, E. Waxman and A. Loeb, *J. Cosm. Astropart. Phys.* **0404**, 006 (2004)

[420] A.W. Strong, I.V. Moskalenko and O. Reimer, *Astrophys. J.* **613**, 956 (2004)

[421] J. Nishimura *et al.*, *Astrophys. J.* **238**, 394 (1980)

[422] F.A. Aharonian *et al.*, *Astropart. Phys.* **17**, 459 (2002)

[423] M.C. Chantell *et al.*, *Phys. Rev. Lett.* **79**, 1805 (1997)

[424] R.J. Gould and G. Schréder, *Phys. Rev. Lett.* **16** 252 (1966)

[425] L. Bergström, J. Edsjö and P. Ullio, *Phys. Rev. Lett.* **87**, 251301 (2001)

[426] Y.B. Zeldovich and I.D. Novikov, *Sov. Astron.* **10**, 602 (1966)

[427] S.W. Hawking, *Mon. Not. R. Astron. Soc.* **152**, 75 (1971)

[428] S.W. Hawking, *Nature* **248**, 30 (1974)
[429] B.J. Carr, *Astrophys. J.* **201**, 1 (1975)
[430] B.J. Carr, *Astrophys. J.* **206**, 8 (1976)
[431] M.Y. Khlopov, B.A. Malomed and Y.B. Zeldovich, *Mon. Not. R. Aston. Soc.* **215**, 575 (1985)
[432] P.D. Nasel'skii and A.G. Polnarëv, *Sov. Astron.* **29**, 487 (1985)
[433] S.D.H. Hsu, *Phys. Lett.* **B251**, 343 (1990)
[434] B.J. Carr and J.E. Lidsey, *Phys. Rev.* **D48**, 543 (1993)
[435] P. Ivanov, P. Naselsky and I. Novikov, *Phys. Rev.* **D50**, 7173 (1994)
[436] J. García-Bellido, A. Linde and D. Wands, *Phys. Rev.* **D54**, 6040 (1996)
[437] J. Yokoyama, *Astron. Astrophys.* **318**, 673 (1997)
[438] M. Crawford and D.N. Schramm, *Nature* **298**, 538 (1982)
[439] H. Kodama, M. Sasaki and K. Sato, *Prog. Theor. Phys.* **68**, 1979 (1982)
[440] S.W. Hawking, I.G. Moss and J.M. Stewart, *Phys. Rev.* **D26**, 2681 (1982)
[441] K. Jedamzik, *Phys. Rev.* **D55**, R5871 (1997)
[442] S.W. Hawking, *Phys. Lett.* **B231**, 237 (1989)
[443] A. Polnarev and R. Zembowicz, *Phys. Rev.* **D43**, 1106 (1991)
[444] J.C. Niemeyer and K. Jedamzik, *Phys. Rev. Lett.* **80**, 5481 (1998)
[445] D.B. Cline and W. Hong, *Astrophys. J.* **401**, L57 (1992)
[446] D.B. Cline, D.A. Sanders and W. Hong, *Astrophys. J.* **486**, 169 (1997)
[447] A.M. Green, *Phys. Rev.* **D65**, 027301 (2002)
[448] M.R.S. Hawkins, *Nature* **366**, 242 (1993)
[449] M.R.S. Hawkins, *Mon. Not. R. Astron. Soc.* **278**, 787 (1996)
[450] E.L. Wright, *Astrophys. J.* **459**, 487 (1996)
[451] D.B. Cline, *Astrophys. J.* **501**, L1 (1998)
[452] A.M. Green and A.R. Liddle, *Phys. Rev.* **D60**, 063509 (1999)
[453] A.M. Green, *Astrophys. J.* **537**, 708 (2000)
[454] B. Carter, *Phys. Rev. Lett.* **33**, 558 (1974)
[455] D.N. Page, *Phys. Rev.* **D14**, 3260 (1976)
[456] D.N. Page, *Phys. Rev.* **D13**, 198 (1976)
[457] J.H. MacGibbon and B.R. Webber, *Phys. Rev.* **D41**, 3052 (1990)
[458] F. Halzen *et al.*, *Nature* **353**, 807 (1991)
[459] D.N. Page and S.W. Hawking, *Astrophys. J.* **206**, 1 (1976)
[460] S.W. Hawking, *Commun. Math. Phys.* **43**, 199 (1975)
[461] J.H. MacGibbon and B.J. Carr, *Astrophys. J.* **371**, 447 (1991)
[462] G.D. Kribs, A.K. Leibovich and I.Z. Rothstein, *Phys. Rev.* **D60**, 103510 (1999)
[463] J.D. Barrow, E.J. Copeland and A.R. Liddle, *Mon. Not. R. Astron. Soc.* **253**, 675 (1991)
[464] J.M. Overduin and P.S. Wesson, *Vistas Astron.* **35**, 439 (1992)
[465] B.J. Carr and J.H. MacGibbon, *Phys. Rep.* **307**, 141 (1998)
[466] D.E. Alexandreas *et al.*, *Phys. Rev. Lett.* **71**, 2524 (1993)
[467] V. Connaughton *et al.*, *Astropart. Phys.* **8**, 178 (1998)
[468] K. Maki, T. Mitsui and S. Orito, *Phys. Rev. Lett.* **76**, 3474 (1996)
[469] J.H. MacGibbon, *Nature* **329**, 308 (1987)
[470] R.J. Adler, P. Chen and D.I. Santiago, *Gen. Rel. Grav.* **33**, 2101 (2001)

[471] J.D. Barrow, E.J. Copeland and A.R. Liddle, *Phys. Rev.* **D46**, 645 (1992)
[472] A.F. Heckler, *Phys. Rev.* **D55**, 480 (1997)
[473] A.F. Heckler, *Phys. Rev. Lett.* **78**, 3430 (1997)
[474] J.M. Cline, M. Mostoslavsky and G. Servant, *Phys. Rev.* **D59**, 063009 (1999)
[475] H.I. Kim, C.H. Lee and J.H. MacGibbon, *Phys. Rev.* **D59**, 063004 (1999)
[476] J.M. Overduin and P.S. Wesson, *Phys. Rep.* **283**, 303 (1997)
[477] P.S. Wesson *et al.*, *Int. J. Mod. Phys.* **A11**, 3247 (1996)
[478] N. Arkani-Hamed, S. Dimopoulos and G. Dvali, *Phys. Lett.* **B429**, 263 (1998)
[479] L. Randall and R. Sundrum, *Phys. Rev. Lett.* **83**, 4690 (1999)
[480] D.J. Gross and M.J. Perry, *Nucl. Phys.* **B226**, 29 (1983)
[481] R.D. Sorkin, *Phys. Rev. Lett.* **51**, 87 (1983)
[482] A. Davidson and D.A. Owen, *Phys. Lett.* **B155**, 247 (1985)
[483] P.S. Wesson and J. Ponce de Leon, *J. Math. Phys.* **33**, 3883 (1992)
[484] H. Liu and P.S. Wesson, *J. Math. Phys.* **33**, 3888 (1992)
[485] P.S. Wesson and J. Ponce de Leon, *Class. Quant. Grav.* **11**, 1341 (1994)
[486] P.S. Wesson, *Astrophys. J.* **420**, L49 (1994)
[487] H. Liu and J.M. Overduin, *Astrophys. J.* **538**, 386 (2000)
[488] J.M. Overduin, *Phys. Rev.* **D62**, 102001 (2000)
[489] S.D. Biller *et al.*, *Phys. Rev. Lett.* **80**, 2992 (1998)
[490] B.J. Carr and M. Sakellariadou, *Astrophys. J.* **516**, 195 (1999)
[491] M.R. Santos, V. Bromm and M. Kamionkowski, *Mon. Not. R. Astron. Soc.* **336**, 1082 (2002)
[492] K. Abazajian, G.M. Fuller and W.H. Tucker, *Astrophys. J.* **562**, 593 (2001)
[493] L. DiLella and K. Zioutas, *Astropart. Phys.* **19**, 145 (2003)
[494] A.G. Doroshkevich *et al.*, *Astrophys. J.* **586**, 709 (2003)
[495] S.H. Hansen and Z. Haiman, *Astrophys. J.* **600**, 26 (2004)
[496] S. Kasuya *et al.*, *Phys. Rev.* **D69**, 023512 (2004)
[497] E. Pierpaoli, *Phys. Rev. Lett.* **92**, 031301 (2004)
[498] X. Chen and M. Kamionkowski, *Phys. Rev.* **D70**, 043502 (2004)
[499] P. Jean *et al.*, *Astron. Astrophys.* **407**, L55 (2003)
[500] C. Boehm *et al.*, *Phys. Rev. Lett.* **92**, 101301 (2004)
[501] D. Hooper and L.-T. Wang, *Phys. Rev.* **D70**, 063506 (2004)
[502] A. de Rujula *et al.*, *Nucl. Phys.* **B333**, 173 (1990)
[503] S. Dodelson, *Phys. Rev.* **D40**, 3252 (1989)
[504] E.W. Kolb, D.J.H. Chung and A. Riotto, in H.V. Klapdor-Kleingrothaus and L. Baudis (eds.), *DARK98: Proceedings of the 2nd International Conference on Dark Matter in Astro and Particle Physics* (Philadelphia: Institute of Physics Press, 1999), p. 592
[505] H. Ziaeepour, *Astropart. Phys.* **16**, 101 (2001)
[506] V. Berezinsky, M. Kachelrieß and A. Vilenkin, *Phys. Rev. Lett.* **79**, 4302 (1997)
[507] I.V.M. Albuquerque, L. Hui and E.W. Kolb, *Phys. Rev.* **D64**, 083504 (2001)
[508] V. Berezinsky, M. Kachelrieß and S. Ostapchenko, *Phys. Rev.* **D65**, 083004 (2002)

[509] H.B. Kim and J.E. Kim, *Phys. Lett.* **B527**, 18 (2002)
[510] E.A. Baltz and L. Bergström, *Phys. Rev.* **D67**, 043516 (2003)
[511] J.L. Feng, A. Rajaraman and F. Takayama, *Phys. Rev. Lett.* **91**, 011302 (2003)
[512] J. Hisano, S. Matsumoto and M.M. Nojiri, *Phys. Rev. Lett.* **92**, 031303 (2004)
[513] N. Arkani-Hamed, S. Dimopoulos and G. Dvali, *Phys. Rev.* **D59**, 086004 (1999)
[514] L.J. Hall and D. Smith, *Phys. Rev.* **D60**, 085008 (1999)
[515] S. Hannestad and G.G. Raffelt, *Phys. Rev. Lett.* **87**, 051301 (2001)
[516] J.A.R. Cembranos, A. Dobado and A.L. Maroto, *Phys. Rev. Lett.* **90**, 241301 (2003)
[517] H.-C. Cheng, J.L. Feng and K.T. Matchev, *Phys. Rev. Lett.* **89**, 211301 (2002)
[518] E.W. Kolb and R. Slansky, *Phys. Lett.* **135B**, 378 (1984)
[519] G. Servant and T.M.P. Tait, *Nucl. Phys.* **B650**, 391 (2003)
[520] G. Bertone, G. Servant and G. Sigl, *Phys. Rev.* **D68**, 044008 (2003)
[521] G. Bertone, D. Hooper and J. Silk, *Phys. Rep.* **405**, 279 (2005)
[522] J. Ellis, J.L. Lopez and D.V. Nanopoulos, *Phys. Lett.* **B247**, 257 (1990)
[523] J. Ellis *et al.*, *Nucl. Phys.* **B373**, 399 (1992)
[524] K. Benakli, J. Ellis and D.V. Nanopoulos, *Phys. Rev.* **D59**, 047301 (1999)
[525] J. Ellis, V.E. Mayes and D.V. Nanopoulos, *Phys. Rev.* **D70**, 075015 (2004)
[526] A.S. Eddington, *The Philosophy of Physical Science* (Cambridge: Cambridge University Press, 1939)
[527] P.S. Wesson, *Observatory* **120**, 59 (2000)

# Index